FOOD
MICROBIOLOGY

FOOD MICROBIOLOGY
A Framework for the Future

By

ANTHONY NELSON SHARPE, B.Sc., Ph.D., F.R.I.C.

Microbiology Automation Section,
Microbiology Research Division, Bureau of Microbial Hazards,
Food Directorate, Health Protection Branch,
Health and Welfare,
Ottawa, Ontario, Canada

CHARLES C THOMAS · PUBLISHER

Springfield · Illinois · U.S.A.

Published and Distributed Throughout the World by

CHARLES C THOMAS • PUBLISHER

Bannerstone House

301-327 East Lawrence Avenue, Springfield, Illinois, U.S.A.

©*1980, by* CHARLES C THOMAS • PUBLISHER

ISBN 0-398-04017-6

Library of Congress Catalog Card Number: 79-24138

With THOMAS BOOKS careful attention is given to all details of
manufacturing and design. It is the Publisher's desire to present books that are
satisfactory as to their physical qualities and artistic possibilities and
appropriate for their particular use. THOMAS BOOKS will be true to those
laws of quality that assure a good name and good will.

Library of Congress Cataloging in Publication Data

Sharpe, Anthony N.
  Food microbiology.

  Includes index.
    1. Food—Microbiology. I. Title.
  QR115.S48   576'.163   79-24138
  ISBN 0-398-04017-6

Printed in the United States of America

C-1

PREFACE

I BEGAN THIS BOOK simply as a personal and somewhat defensive view of the possibilities of providing food microbiologists with mechanical aids, instruments, or automated methods of analysis. More than a century after the beginnings of the science, the poor microbiologist, unlike his colleagues in many newer sciences, has precious little to choose from. As the book developed, it became apparent that the efforts of scientists and engineers to develop these instruments have, for a long time, been frustrated by the very concepts on which food microbiology, its analyses, and standards are based. Worse still, it now seems to me that the emphasis placed on these same concepts may have navigated food microbiology into a scientific backwater in which it could mill around for many years. Much research is carried out on microorganisms and foods; we progress, of course, but to where? Food microbiology boasts analytical methods that in their exquisite sensitivity and selectivity are almost without equal in science. And yet, we would be hard pressed to assure one consumer whether or not his sandwich will make him ill, now or in ten hours time. Could the reason be the overwhelming but quite unnecessary importance we seem to attach to microorganisms and their numbers? I hope to show so.

I believe this can be a precise, progressive, and instrumented science but not as the science we know today. Progress and instruments are inextricably linked; if we want to move forward, we must clear microorganisms, Petri dishes, and numerical standards out of the way to make a path for instrumentation. This book, then, offers what may be seen as a rational alternative framework for food microbiology and microbiological analysis. By adopting it we could have instruments almost immediately. In so doing, we could open the door to data with a precision and meaningfulness unobtainable today.

Literature citations are made only where they seemed essential, for this is not intended as a reference work but as a basis for discussion. Being a personal view, it is subjective, biased, and prejudiced. I hope it will be controversial, for the subject is in need of stirring up.

INTRODUCTION

MOST OF US TODAY live fuller and healthier lives than our great-great-grandparents could have dared hope for. In no small measure, the standard of health we enjoy derives from the wholesomeness or quality of our food. And this, in turn, derives directly from the efforts of several generations of microbiologists who undertook to concern themselves with food, and of lawyers, civil servants, politicians, consumer's associations, journalists, and food manufacturers who, through altruism or otherwise, also busied themselves thus. The protection afforded us by this army of specialists is fortified by standards, guidelines, and other specifications for the microbiological quality of foods, and by an arsenal of recommended methods for the examination of foods. We live well behind this protection; to many of us the thought of food poisoning is no more than a subject for snide jokes.

It may seem odd, therefore, that having admitted this I now devote a book to a proposal to abandon many of the concepts on which our well-being rests. To do so implies a dissatisfaction with the system and a belief that things could be better. It is certainly not a proposal one makes lightly, ignoring the rush of criticism that may follow. The existing system has served us well; undoubtedly it could continue to do so for many years. We understand it, we trust it and, as in all matters relating to human health, caution — even conservatism — must be the rule. A new system is not likely to be readily accepted unless it has many attractions over the old.

Indeed, I am not that unhappy with the existing system, and I put forward the case that follows with far more tongue in cheek than opiniatry. At the worst, it will be ignored. At best, it may provoke some reexamination of the objectives of food microbiology and microbiological analysis, and there are few established thought patterns that do not benefit from a little introspection.

This old hulk of a science has survived, little changed, for more than a century. If its frames are still sound, a little poking around will do no harm. If it just happens that we find it a trifle shaky under the accumulated weight of 110 years of observations, ambitions, and conflicts, it will be useful to have an alternative structure around in which to take refuge.

It may be relevant to the argument that follows that I am not a microbiologist by training. To some extent, my interests still stand remote from the mainstream of the science, and there must surely be times when any discipline needs the views of those who do not see problems through the established screen of training. It is a matter of conditioning, and as scientists we are no more immune than others to its effects. In my experience, at least, it is interesting that persons from other disciplines, on making contact with microbiology, generally seem more open to facile solutions than those formally trained in the subject. I remember, for example, a chemist who, having acquired the responsibility of his company's microbiological quality control laboratory, experienced difficulties in maintaining a supply of sterile diluent for the plate counts. He solved the problem neatly by adding a little thiosulphate solution to the London tapwater and used it directly as the diluent, to the horror of his peers who maintained that (a) all materials used for counts should be sterile, not "nearly sterile," and (b) without added salt or peptone, bacteria are known to lose viability.

For this and similar reasons, his data thereafter served as a source of suspicion and amusement. In truth, though, the newcomer had a satisfactory solution to his problem, for the level of contamination in the tapwater was insignificantly small compared with the levels counted in the plates; he had destroyed excess chlorine in the water by the thiosulphate treatment, and the time taken for diluting the food before the bacteria were immersed in the isotonic growth medium was so short that an insignificantly small loss of viability occurred.

By this I do not mean to suggest that seeing problems in a different light necessarily means the light is better. The aphorism, "If you can keep your head while others around you are losing theirs, maybe you don't understand the situation," was never truer than in microbiology. The potential for the nonmicrobiologist to make useless developments through misunderstanding or ignorance of the ramifications of the subject is enormous. A long while ago, several other colleagues and I, newly introduced to microbiology from the "harder" sciences, were quite convinced that determination of physical data for antibacterials would clear up the matter of their modes of action and lead to powerful new compounds through extrapolation and prediction. This seemed particularly so when, for example, adsorption isotherms for tribromosalicylanilide (TBS) indicated that at its minimum inhibitory concentration, the quantity adsorbed on *Straphylococcus aureus* would just coat the entire surface of the cell with a monomolecular layer. The mechanism seemed so straightforward — TBS just blocked up all the holes. Now, many years later, neither they nor I would expect bacteria to behave quite as docilely and passively as particles of graphite, and doubtless we all remember the tolerance our microbiological and biochemical colleagues afforded us. Mark Twain once said, "When I was a boy of fourteen, my father was so ignorant I could hardly stand to have the old man around. But when I got to be twenty-one, I was astonished at how much the old man had learned in those seven years. . . ." I have grown up a little myself over the years, but microbiology is a big subject. I think I see a simpler path through its maze; I sincerely hope posterity will not show me to still be seeing the subject in the dim light of ignorance.

In setting the theme for the following chapters, it will be useful to describe an important aspect of the problem. That this is the overwhelming importance ascribed to microorganisms may seem illogical when the subject under discussion is called food microbiology; however, the reason should slowly become apparent.

ACKNOWLEDGMENTS

HOW DO YOU ACKNOWLEDGE everyone who by providing information, help, or inspiration contributes to the fashioning of your opus? Or rather, where do you draw the line in naming them? I have gained from too many, in Health Protection Branch and in Unilever's Colworth House research laboratory in particular, but elsewhere as well. To all of these I say — thank you. Most of you will recognize your personal contributions. Some of you may have wondered how I managed to know so little at times.

Just three personal acknowledgments. To Dr. Hilliard Pivnick and Dr. André Hurst for help and encouragement in more ways than they might even care to recognize, and to Dr. David S. Clark, for maintaining an environment in which lateral thinking about the establishment, as exemplified by this book, can flourish.

A.N.S.

CONTENTS

FOOD
MICROBIOLOGY

Chapter 1

ON FOOD MICROBIOLOGY

FOOD MICROBIOLOGY is the name given to the science dealing with microorganisms in food, the means by which they enter it, their ability to multiply in it and produce undesirable substances or survive to infect us, and with methods of analyzing foods to detect them. The objectives of persons concerned with it vary. On the regulatory side are those whose aim is to protect the public from food borne microbial hazards and otherwise ensure that food that is sold is of acceptable quality. On the other side are those whose main interest is ensuring the maximum commercial exploitation of materials without running afoul of either regulatory agencies or adverse publicity. The probable behaviors of microorganisms are of great interest to both sides. Microorganisms have been cultured in tubes, in Petri dishes, on filters, in chemostats, in broths, jellies, and emulsions; counted, purified, identified, characterized, and preserved. Careers have been devoted to searching for conditions most favorable to their growth, describing the dynamics of their populations and the means by which they can be injured or killed.

In terms of the scientist time invested in them, then, food microbiology serves microorganisms very well. We are, of course, quite pleased for the microorganisms, although that should not stop us asking how well it serves human beings.

In the three years from 1973 through 1975, among the population of Canada, the USA, and England/Wales, 2,461 outbreaks of food poisoning involving a total of 42,524 persons and 189 deaths were officially reported as having been caused by microorganisms. In Japan alone, from 1968 to 1972, 3,217 outbreaks involving 119,897 persons were recorded.[1] In Canada and the USA, the largest percentages of cases were due to infection by *Salmonella* species, followed closely by poisoning from staphylococcal toxins, and then varying percentages of illnesses from *Shigella* species, *Clostridium perfringens*, *Clostridium*

3

botulinum, Bacillus species, and *Vibrio parahaemolyticus.* In Japan, reflecting the importance of marine products in the diet, the greatest percentage of cases was caused by *Vibrio parahaemolyticus.* In addition to this, some proportion of 34,098 other cases of food poisoning in the three western countries, and 58,022 in Japan, which were recorded as being of unknown etiology, may have been caused by microorganisms during the same years. These figures are not enormous compared with the populations of the countries concerned, but it is obvious even from this that small cracks still exist in our system of food quality protection.

By no means are all illnesses officially recorded, however, and epidemiological evidence indicates that the true figures may be very much higher. In Canada alone, for example, microbiological contamination is suspected of causing as many as 400,000 illnesses a year, burdening the country with costs of at least $100,000,000 in medical services and lost productivity.[2] On the global scale, it has been estimated that on any given day of the year, some 200,000,000 human beings are afflicted with acute gastroenteritis.[3]

Against this seeming failure in protection should be weighed an impressive activity by regulatory agencies in preventing potentially hazardous food from reaching consumers in these countries. Publications such as *Protection* (Quarterly Report of Enforcement Activities, Health and Welfare, Canada) and the *FDA Consumer* (the official magazine of the Food and Drug Administration) contain reassuring lists of recalls, seizures, shipments refused entry, etc., on the grounds of either demonstrated presence of microbial contamination, through rodent or insect infestation or similar filth, or other evidence of insanitary handling, all of which imply the possibility of serious microbial contamination. The *FDA Consumer,* for example, from December 1975 to November 1976, lists 157 seizures of food lots on grounds relating to microbial contamination. Rodent and insect infestation or other evidences of having been prepared, held, or packed under insanitary conditions account for the majority (119) of these, with a minority attributed to more direct evidence such as decomposition (19), swollen cans or lids (12), presence of mold (9), *Escherichia coli* or other coliforms (6), *Salmonella* (1) and aflatoxin (1). The magazine *Protection,* during 1977, listed 27 recalls, 25 seizures of products, 61 shipments refused entry into Canada, and 12 convictions for offences under the Food and Drugs Acts relating to microbiological quality. A total of 56 items referred to infestation or insanitary handling, 22 to contamination with *Salmonella,* 6 to other evidences of microbial contamination, and 25 to evidences of inadequate packing (usually faulty cans) or underprocessing.

Obviously, the viewpoint one takes of the efficiency of food microbiology in protecting us will depend on one's biases. Protection is, by its very nature, an undramatic service, noticeable mainly when we are reminded from time to time about illnesses we do not get, and difficult to quantify other than by the dubious method of pointing to the ravages occurring in less fortunate populations. It is good to be protected in health but also in our wallets. Human nature being what it is, we may sometimes question the price we pay for the protection received. During 1977, foods seized in Canada through evidence of infestation or overt microbial contamination amounted to a value of only $86,000. Typical listings in *Protection* read

Peanuts	All 16 oz stock manufactured from Jan. to end of Mar. 1977.	Aflatoxin
Green beans	All lots produced in summer 1976	Defective cans
Salami and sausage	All stock	*Salmonella* contamination
Hearts of palm	28,800 cartons of 230 g	Elevated pH
Frog legs	3,945 kg	*Salmonella* contamination

It is difficult to find accurate data, but the value of food destroyed or diverted as a secondary result of regulatory activities must be enormous. The items described officially probably represent a minute fraction of the total value of food voluntarily recalled or diverted from the country in anticipation of official curiosity. Likewise, the quantity of food wasted through spoilage in homes or on the shelves of grocery stores can probably only be guessed at.

Regulatory agencies, public health departments, and other

organizations responsible for our safety carry out their missions of food quality surveillance and enforcement superbly, within the means at their disposal. Nobody would seriously suggest that others could do a better job with the tools available. What may be questioned, however, is whether or not these well-tried and tested tools — that is, our methods for analyzing foods to determine whether the existence of microorganisms in them is likely to make them unduly hazardous or unwholesome — will be seen to be the best available in, say, the twenty-first century.

Are those tools adequate even for today? Certainly, a great deal of research on methods of detecting microorganisms has already been done and is continuing. The existing analytical methods are already extremely sensitive and selective. Indeed, in Chapter 3 I shall show that these old, established methods are unlikely to be surpassed by any methods derived from the more "advanced" sciences in the foreseeable future. What must be examined is the value of assessing the quality of foods on the basis of the microorganisms we can detect in them rather than on, say, their capacity to evoke symptoms of disgust or illness in those who encounter them. The difference may seem trifling since we know microorganisms are very pertinent to quality. At times, certainly, it is vanishingly small. Nevertheless, approaching the subject from this slightly different angle can sometimes provide us with a very different perspective, and the implications are far-reaching, as will be seen.

Within this difference, I believe, rests any possibility we have for increasing the efficiency of our control over the microbiological quality of food. Within it rests the possibility of food microbiology ever becoming a heavily instrumented science. And within it also are hidden statutory standards and data, direct and absolute, to a degree unobtainable today.

It is worth asking where our concern with microorganisms in food has led us. We know a great deal about the activities and properties of many microorganisms already, as pure laboratory cultures. We have isolated, classified, and characterized many species, strains, and variants. But putting names to them does not mean that we really understand them or always know what they will do in the context of a food. To be able to do that requires a great deal more knowledge than we currently possess. And when our knowledge is inadequate we rely on generalizations. Such things are fine for the lucky majority who fall within their limits; they are a bane, however, for those unfortunate enough to fall outside their fringes. Such people suffer the cramps, diarrhea, or worse.

Can we, for example, assure one person whether or not his mouthful of food will make him ill? Is it likely that continuation of research on methods of detecting microorganisms will lead us to a stage where we *can* make such an assurance? The answer to both questions must generally be *no*. The only assurances we can make — and are likely to be able to make — are of a very broad, statistical nature. Thus, we are confident that lowering standards, that is, permitting higher levels of various organisms in foods, would lead to an increase in incidences of food poisoning across the North American continent. The blanket of statistical probabilities smooths out point to point variations over large populations and hides the inadequacy of our knowledge of behaviors in individual situations. We do the best we can, but the sufferer from food poisoning can be forgiven for gaining little comfort from the statistics of the matter. He is interested only in his individual circumstance, and it is there that our approach to microbiology in food breaks down.

We can be fairly sure that during one phase of its growth a culture of *Staphylococcus aureus*, strain S6, will double its numbers every 17 minutes when growing in trypticase soy broth at 35°C. We can be fairly sure that strain 361 of this organism growing at 35°C in a broth containing 3 percent protein hydrolysate, 3 percent NZ-amine-type NAK, 0.00005 percent thiamine, and 0.001 percent niacin at pH 6.7 can be induced to produce 100 to 150 μg/ml of enterotoxin C_2 in 24 hours.[4] We are much less sure how much enterotoxin C_2 this organism will produce in, say, the environment of a ham and lettuce sandwich, where it is exposed to varying conditions of temperature, pH, aerobicity, nutrient supply, etc., and where it must compete or cohabit with other organisms. We can be fairly sure that enterotoxin C_2 will cause vomiting and diarrhea in 50 percent of cynomolgus monkeys at a level of 0.04 μg/kg animal weight. We are sure that it can be a causative agent in staphylococcal food poisoning in humans, along with several other toxins, but we are much less sure of the relation between dose and response for ourselves than we are for cynomolgus monkeys. We are quite sure that there is an enormous range in the abilities of the hundreds of known strains of *Staphylococcus aureus* to produce enterotoxins, just as we are sure that there are many more strains yet to be discovered, along with many more enterotoxins. When we weigh the unknowns against the knowns, we see that the odds against us being able to assure the eater that his ham and lettuce sandwich will be fit to eat in ten minutes or ten hours time are very small. With the present capability of the traditional approach, we could easily count the number of dispersible *Staphylococcus aureus* groups in the sandwich. With a little more effort, we could count the number of those having the quantized* property of coagulase production. That there exists a satisfactory correlation between quantized coagulase activity and quantized enterotoxin production, we could take on trust from the literature. From this point, we could only guess at whether the sandwich is capable of causing illness in this eater, for we really have little information about the ability of the organisms we have counted to survive, multiply, or produce enterotoxin in this individual situation.

Moreover, if we are going to retain our preoccupation with the microorganisms, making such assurances requires so much more information than we are currently able to provide that it would seem to be an impossible achievement in the foreseeable progress of science. This is a great source of frustration for many microbiologists; had we been interested in mercury contamination in the sandwich, for example, a few moments work with an atomic absorption spectrometer would have provided us with an answer.

This gloomy picture should be lightened with the qualification

---

* Coagulase, enterotoxin, and many other biochemical activities tend to be quantized by microbiologists through the assignment of + or − characters, depending on subjective assessments of the presence or absence of activity. This nomenclature conveys little information about the level of activity, save perhaps that implied in assignments such as ‡ or ‡. In any case, observations of these activities are usually made under growth conditions differing greatly from those in the food, and such changes are known to have profound effects on many of these activities.

that methods are now becoming available for, say, detecting some toxins directly at levels thought to be physiologically active. Thus, it is now becoming possible, as in the case of mercury, to measure a parameter of much more immediate importance to the consumer. This takes us a large step closer to the proposal in this book, namely, that the quality of foods be assessed on the basis of their ability to generate responses in humans. However, for as long as official standards, guidelines, etc. for *S. aureus* and other organisms continue to be written in terms of their countable numbers at the instant of sampling, the true value of these more direct methods cannot be realized.

The example just given for *S. aureus* is typical of situations existing in all aspects of food microbiology, from quality control in factories to the tracing of sources in outbreaks of food poisoning. In all cases, the traditional microorganism oriented approach leads to data whose main value is statistical rather than immediate and useful in the overall control of wholesomeness rather than in individual cases. The microbiologist in a meat products factory, for example, may know that the general level of counts in his products or at critical points in the plant may be higher in summer than in winter. He may, through experience, be sure that if his methods of analysis remain unchanged, few problems (complaints, recalls, etc.) can be expected from a product as long as levels of microorganisms do not rise above a certain value. But it is a statistical effect; he cannot be sure whether or not the level of microorganisms in any individual product lot predisposes that lot to a short shelf life. He may count *Micrococcus* and *Leuconostoc* in weiners before they leave the plant. But weiners are apt to spoil through development of surface slime or souring, and their shelf life before these undesirable effects are manifest bears little relation to the levels of these organisms before the product is distributed. Notwithstanding, only allowing weiners to be distributed when the count is less than, say, 10^6/g is statistically likely to result in a relatively trouble-free product.

What, however, is the microbiologist to conclude about a production lot that seemed satisfactory but that resulted in large numbers of complaints? What failed? Was it the analysis? Was the overall level of microorganisms actually much higher than indicated? Did the product suffer some undetected abuse, e.g. incorrect cabinet temperature, in the store? Or was the microbial flora unusually active at producing slime or acid?

Similarly, what is he to conclude about a manufacturing lot when its count just exceeds the arbitrary limit he has set through experience? Should he allow it to be distributed? Certainly the plant manager, who is committed to shifting products through the plant, will question the value of this limit if it means diverting or destroying the product. Can the microbiologist say for sure that it will create complaints, adverse publicity, and loss of sales by the company to a value greater than that of the lot itself? He cannot, for there are too many unknowns. His dilemma was set when interests became centered on an individual case and — as will be reiterated many times in this book — it is here that enumerative microbiology must be seen to be inadequate.

Many alternative techniques to counting microorganisms are available. Some of them are very relevant to the information required. However, they tend to be little used through fear that the data they provide will not ensure satisfactory protection against contrary findings by regulatory agencies, who generally rely on the traditional analyses embodied in official or recommended methods of analysis. The existence of standards for the microbiological quality of food thus tends to lock all food microbiology into the accepted format and perpetuates occurrences of the kind of dilemma described. While it is doubtful whether the quality of food could be controlled effectively across a country without statutory standards of some kind, we have now reached a state in the development of sciences relating to microbiology where alternative and more pertinent measurements are possible. We should be looking critically at the level of efficiency our existing standards, through the aged concepts embodied in them, impose on the science as a whole. Certainly, the continued existence of standards based on microbial numbers will ensure that (a) food will frequently be wasted on the grounds of violating statutory standards when, in fact, it may have been quite consumable, and (b) food poisonings will persist whenever combinations of microorganisms and growth conditions occur that are outside the statistically most probable range.

Perhaps the first (false positives) would be quite acceptable if the second (false negatives) could thereby be eliminated. However, at the moment we live between the two.

These standards may lead to paradoxes of a kind the ancient Greeks might have loved. Paradoxes, that is, resulting from ambiguities of data in which good food may be condemned by authorities (who are more or less bound by the statutes) to the detriment of the very populations they were set up to protect. In large measure, these are simply interpretation difficulties, resulting from the unfortunate tendency of our present standards to relate to the instant samples arrive in the laboratory rather than to the instant the food is consumed. Significantly, in attempting to provide more meaningful data within the framework they impose upon us, we seem to be dooming ourselves to an eternity of unnecessarily so and ever more laborious microbiological analyses.* The alternative may be to ask whether our standards adequately describe what we seek to define about our food.

You may believe it or not that a school of microbiology was running as early as 450 B.C. in the Gymnasium of Athens. By that time, all of the philosophical principles of the growth of bacteria and their effects on man and the environment had been deduced, debated, and described under the guidance and tutelage of the venerable bacteriologist, Inquisitus. By 390 B.C., the Senate had laid down standards for the microbiological quality of all foods, with the object of protecting the health of the Greek people. Strictest enforcement was observed over the organism *Bullus fecalis* since it was well known that eating foods containing this organism resulted in a high proportion of fatalities.

In 317 B.C., a great famine fell upon Greece. Some frozen Carthaginian goat's meat was captured and brought to the Gynmasium for examination, whereupon Inquisitus decided that his two best students, Orthodoxus and Hereticus, should determine if it was fit to eat. The two set to work and, two days later, presented their findings to the hungry Inquisitus. Hereticus pleaded that the meat was fit to eat, while Orthodoxus warned that to eat it would be fatal. A debate then ensued.

"See," said Orthodoxus, waving his laboratory notebook, "I weighed a piece of the meat, blended it in sterile diluent, made several serial

---

* The occasional development of ingenious shortcut techniques notwithstanding.

dilutions of it in bottles, inoculated some dishes and added sterile agar jelly, then incubated the dishes in an atmosphere of hydrogen. Later, I counted the colonies that had formed and that seemed to me to resemble those of *Bullus fecalis,* then subcultured some of them and performed a series of biochemical tests so as to confirm that a proportion of them were, indeed, this organism. This is the method we call the plate count. By this means I determined that the goat's meat contained 173 cells of *B. fecalis* per scruple, and this is in gross violation of the limit laid down by Senate. Since the meat has been warming steadily since its capture, the numbers of *B. fecalis* in it will certainly have increased still further, and it will kill us."

"The plate count does not lie," said Inquisitus solemnly. "It is regrettable, but we must destroy the meat, though our stomachs cry for sustenance."

"To do so would be foolish," said Hereticus. "The meat is quite safe and, in any case, we should surely perish in the famine."

"How can you say that, Hereticus, when you spent so little time over your analysis?" asked Orthodoxus. For Inquisitus' benefit he added, "I could not help noticing that you were asleep in the sun the whole time I was working in the laboratory."

"Oh no!" Hereticus yawned, "that is not quite true. I did incubate a piece of the meat and examined it both days for the toxin of *B. fecalis* using my spectroscope. I found none. Therefore, I say the meat is safe."

Inquisitus had at once pricked up his ears. "How can this paradox be?" he asked. "Did you not, Orthodoxus, also test this meat with your spectroscope?"

"The spectroscope is not able to detect toxin from only 173 cells of *B. fecalis,* as you yourself well know, Master," Orthodoxus reminded him.

"That is true," Inquisitus mused. "How then, Hereticus, can you say that this meat does not violate our standard?"

"I did not claim that," Hereticus replied. "But you know we are able to detect toxin with the spectroscope when it has reached a dangerous concentration. Therefore, I incubated my sample of the meat so as to allow the bacteria to multiply if they could. Even then I did not detect the toxin. Thus, I repeat that the meat is safe."

Inquisitus scratched his head. "Do you doubt the evidence of Orthodoxus that the meat contains *B. fecalis?*"

"Not in the least," said Hereticus. Turning to Orthodoxus he asked, "But, Orthodoxus, did you also determine the rate at which *B. fecalis* cells are able to multiply in the meat? We know there are great variations in the growth rate between the strains of all organisms."

"I have no data on this," admitted Orthodoxus.

"And also," continued Hereticus, "While you were so busy in the laboratory, did you determine the rate at which cells of this particular strain of *B. fecalis* are able to produce the toxin? We know there are even greater variations in this property."

"I have no data on this either," admitted Orthodoxus.

Hereticus pressed his point. "And furthermore, Orthodoxus, did you determine what other organisms were in the meat and how many of each there were?"

"I did manage to find time to make the analysis we call the total plate count, and found there were many such organisms, but of what species I cannot say," Orthodoxus replied testily. "However, there were definitely 173 cells of *B. fecalis* per scruple in the meat and that alone is grounds for its condemnation."

"But is it not known, Orthodoxus, that the growth of other organisms may greatly affect not only the growth of *B. fecalis* but also the rate at which it produces toxin?"

"That may be so," started Orthodoxus, "but. . . ."

Hereticus interrupted. "But did you also make other analyses on the meat to determine whether or not it contained inhibitors or promoters of the growth of *B. fecalis*?"

"Of course not!" Orthodoxus replied hotly. "By working so laboriously, I was barely able to complete my plate count on this accursed meat."

"In that case, Orthodoxus," inquired Hereticus, "bearing in mind that you know so little about the properties of this meat and the organisms in it, how can you be sure that it is unfit to eat?"

Inquisitus cut in. "I must interrupt here to assist Orthodoxus, who has labored so hard on our behalf. It is true, Hereticus, that there are many things unknown about this goat's meat. But we are certain of one thing — namely, that the meat contained 173 cells of *B. fecalis* per scruple at the time Orthodoxus began his plate count. Also, we know that even one cell is capable of multiplying to give us a lethal dose and cannot be tolerated in food. Therefore, we must condemn the meat in order to protect ourselves and others who may consume it."

"I agree with your reasoning," said Hereticus respectfully. "But I do not agree with your conclusion since I have tested the meat myself, as you well know, and found that it does not produce toxin. Therefore, I say that although I do not know whether all goat's meat is incapable of producing the toxin, my unknowns are no greater than those of Orthodoxus. And at least in the case of this particular meat, my simple experiment gives me a result that is the true sum of all the variables. I cannot say *why* this particular meat does not produce toxin, but I observe that it does not, and I repeat that we should eat it since it is safe."

Orthodoxus and Inquisitus both descended on him roundly. "But to allow such meat to be eaten violates the regulations that have been passed by the Senate, namely, that one shall not pass for sale or consumption goat's meat containing *B. fecalis.* . . ."

The debate continued thus for some time. It would have continued for much longer, but Orthodoxus and Inquisitus both died of starvation, unlike Hereticus who had secretly feasted on the meat during the night. After dining well for several days, Hereticus gracefully accepted the recently vacated position of Head of the School.

REFERENCES

1. Todd, E. C. D.: Foodborne disease in six countries — a comparison. *J Food Protect. 41:*559, 1978.
2. Advisory Committee on Food Safety Assessment: *Report to the Minister of National Health and Welfare and Minister Responsible for the Status of Women.* Department of National Health and Welfare, Ottawa, Ontario.
3. Gorbach, S. L.: Acute diarrhoea-atoxin disease. *New Engl J Med 283:*44, 1970.
4. Robern, H., Stavric, S. and Dickie, N.: The application of QAE-Sephadex for the purification of two staphylococcal enterotoxins. 1. Purification of enterotoxin C_2. *Biochim Biophys Acta 393:*148, 1975.

Chapter 2

A VICIOUS CIRCLE

THE UGLY DUCKLING

To CALL INTO QUESTION, however slight the grounds, the efficiency of food microbiology at protecting society from the effects of microbiological unwholesomeness, is likely to provoke the immediate and very understandable response that if instrument manufacturers or other researchers like myself would come up with practical instruments for microbiological analysis, then efficiency would at once improve. It is very obvious to all food microbiologists, for example, that the number of analyses they are able to make with the time and equipment available to them is far less than the number they consider desirable if they are to adequately monitor foods.

That this is a very valid point will be seen when it is considered that a well-equipped and programmed regulatory laboratory serving a large city will do well to handle more than about 1,250 subspecimens per active microbiologist member per year.* On the average, each subspecimen may require four analyses, e.g. standard plate count, coliforms, *Staphylococcus aureus,* and *Salmonella,* a total of 5,000 analyses per microbiologist. The cost of running the microbiological side of the operation alone may prohibit the employment of more than five to ten active microbiologists when account is taken of the laboratory support staff (media and glassware preparation, etc.) and field operatives (inspectors, advisors, etc.) that are also required. In industry, 50 or 100 analyses per day, distributed between products, machinery, and personnel, may have to suffice in a food manufacturing plant turning out several hundred tons of food a day, in 20 or more different product lines.

---

* This figure is somewhat higher than typical outputs suggested by management and staff resource utilization publications such as the *Canadian Schedule of Unit Values for Clinical Laboratory Procedures* (Statistics Canada). It is a practical figure, reflecting the efficiency of the laboratory from which it was taken.

There is no doubt that regardless of how efficient the actual analytical methods used are, an analytical capability spread so thinly across the enormous quantity of raw materials, machinery, and products manufactured, distributed, stored, or handled in retail catering establishments or homes is bound to profit by simply increasing the number of analyses carried out. The picture we can currently obtain is very incomplete. Most microbiologists subscribe to the general clamor for mechanization, instrumentation, or automation.* At least one factor in their reasoning is the desire to obtain a more effective picture of microbial contamination by analyzing greater numbers of samples.

The future efficiency of food microbiology is inextricably bound to the development of instruments. However, while most microbiologists would expect to see developments that provide them with larger quantities of conventional data, i.e. microbial counts, it may well be that the arrival of these instruments initiates a radical change of attitude to and requirements for the whole subject. Just as the horse was displaced as a significant transport factor by a concept more commercially practical than the mechanical leg, it is likely that realistic future instrument developments for food microbiology will be based on something other than the mechanical or simulated plate count.

We look to increased productivity, as for example, in samples analyzed per day, as a major benefit from instrumentation. It is odd, therefore, that despite an obvious and acknowledged need for it, an insignificant level of instrumentation has so far appeared in food microbiology laboratories. Most microbiologists today carry out their analyses by hand, using methods that have changed little over the last 100 years.

Indeed, to anyone familiar with food microbiology, developments in mechanization or automation appear to have almost reached an impasse. Compared with other branches of science such as physics or chemistry — even with clinical microbiology —

---

* These three words are often used interchangeably in papers relating to food microbiology. While *instrumentation* can be used to cover all circumstances, *mechanization* should strictly be used to describe the use of mechanical aids in part or parts of analytical processes, leaving *automation* to describe processes carried out without human control between entering samples and recording data.

the range of instrumental aids available to food microbiologists is appallingly small. Quantitative food microbiology has stagnated for many years in this regard, while all other sciences moved ahead. Ask any physicist or chemist what the exciting developments in their fields are, and they will point to any one of hundreds, ranging over accelerators, spectrometers, gravity wave detectors, radiotelescopes, high pressure liquid chromatographs, probe microanalyzers, and many more. Ask the same question of a food microbiologist, and his answer may well be combined selenite/tetrathionate broths for *Salmonella* enrichment. . . . Indeed, the most significant advance in the efficiency of food microbiology this century has almost certainly been development of the disposable plastic Petri dish.

The disparity in technology has existed for so long that food microbiologists are in danger of becoming resigned to the situation. And to some extent they might well do so, for we should certainly ask very pointedly whether, for example, the introduction of automated methods is even desirable if there is the slightest chance of it resulting in decreased employment opportunities for microbiologists. However, at the moment, the majority of food microbiologists would argue enthusiastically for the introduction of gadgets, mechanical aids, machines, instrumental aids, or automated instruments into their laboratories if such devices were shown to be available, affordable, and demonstrably preferable to human technicians through their sensitivity, accuracy, reliability, analysis times, throughput, safety, space requirements, running costs, cost per analysis, readiness, versatility, adaptability, awareness, judgement, credibility, and so on. The matter is not so clear-cut however, as to be merely a shortage of supply in the face of explicit demand. Some of these factors are judged on very subjective grounds. Blame for the scarcity of instruments does not rest entirely with instrument manufacturing companies but with ourselves. Whether we will actually drop traditional methods in favor of such devices when somebody surprises us by coming up with an instrument that really works is open to question and is a problem about which much of Chapter 7 revolves. When an instrument company sends around its questionnaire, to put down that we would be keen to have the instrument is one thing, but if the crunch comes, the thought of having to (a) go to the boss with a case for spending money and then (b) jeopardize our reputation by actually using this new technique is apt to make us peer much more closely at the realities of the method, and its apparent attractiveness may change quickly. Faced with a choice of carrying on in the old familiar manner or sticking our necks out into an immature technology, it is easy to be too conservative, particularly so when we can instantly justify ourselves by pointing out the unknowns involved and their possible disastrous consequences in our responsibility to the public. We are accustomed to being the sole judges in matters relating to our science, for there are enough ramifications and elements of witchcraft about microbiology to always give us an edge over outsiders who would question our decisions.

Nevertheless, there is an accepted need for much greater levels of instrumentation in food microbiology. There is no doubt that development of satisfactory and successful instruments is the key to the future of the science. Thus, we should place the greatest importance on determining why food microbiology continues to live in the Dark Ages where the worth of a man is measured by the skill in his hands and the sharpness of his eyes — an idyllic situation but one fast becoming economically untenable.

Many microbiology laboratories have made their own inventions to speed or simplify routine operations: devices for holding autoclavable wastes, agar melters, multipoint inoculators, timers on incubators, grids and screens for counting colonies, and so on. These are personal things, not taken up by laboratories in general, and though they may work excellently in the hands of their inventors, they cannot be called successful. To be classed as such, an invention must be able to be found in large numbers of laboratories. That is, it must be made commercially; there is no other way inventions of sufficient complexity to be significantly useful can find their way into wide-scale application. An invention is only successful when a commercial company has invested sufficient money, time, and effort in developing it as a marketable product, advertising, exhibiting, and distributing it, and when it has been bought by enough laboratories for the company to realize an adequate profit on its investment. An instrument that is thus successful will see the company coming back to its inventors for more. If not, the company will consider that it has burned its fingers and will turn to more profitable areas next time. If those areas do not include food microbiology, we are the losers, and realization of the full potential of the science is delayed yet again.

I believe that unless many of the concepts of food microbiology change considerably during the next few years, there is very little chance of practical (accurate, reliable, versatile, speedy, easy to use, etc.) instruments being developed and marketed to the multitude of food microbiology laboratories, in particular, to the quality control laboratories of small food manufacturers, who may be the most needy. This will be due to a lack of interest — a reluctance to provide the necessary investment — on the part of commercial scientific instrument companies. But no one should blame them. They are in business to make money for their shareholders, and food microbiology is a very poor subject for any Board to speculate with for their shareholders' approval.

If we are ever going to furnish food microbiology laboratories with the instruments we think we need so that we do not look like backward relatives, we have to make ourselves more attractive to the larger instrument manufacturing companies. Only they have the resources to develop our instruments and get them on the market within a reasonable period of time. But how does the ugly duckling go about turning into a swan? If we had the right genetic material, nature would do it for us in the fulness of time. I don't think we do though, and I think we should seriously consider some cosmetic surgery. And as we discuss the form our new features could take, we may also see how our discipline could, at the same time, be made more scientific, more meaningful, and of greater service to the public. True beauty is functional, and we are going to discuss how to make microbiology beautiful. But first, we must identify our shortcomings.

THERE *ARE* PROBLEMS

An easily visible area of business does not drift along for years without being exploited unless it presents unusual technical, financial, or other difficulties. A sales potential for instruments in food microbiology must certainly be visible to the 6,000 or more instrument manufacturing companies in North America,* even if it does not have the glamour or apparent rewards of, say, transport or energy. Few data are available regarding the potential market field. However, at least 270,000 people in North America may be exposed through their training to food or clinical microbiological analysis at some time or other,† although only a small proportion of these will actually practice the subject. Approximately 32,000 subscribe to the major North American microbiology based scientific journals, for example. Bearing in mind the general expensiveness of scientific instruments, these people alone would seem to represent an appreciable market field.

There are also at least 775,000 registered establishments (hotels, food stores, eating and drinking places, etc.) in North America† dealing with food in some way. If only a very small fraction of these saw benefits from having an instrumental watchdog on their premises, another large market might be defined.

Most food microbiologists will agree that the field has not been satisfactorily exploited by scientific instrument manufacturers. There have to be reasons for this, and we should find them. If we describe the difficulties adequately and, as a scientific community, persuade ourselves to give them proper recognition, there is a possibility of doing something about the situation. It is no use simply being aware that something is wrong; we need to know just *what* stifles development investment in food microbiology. Do the problems lie with industry, with food microbiologists, or are they completely external? Only a recognition of the problems is likely to stimulate those involved into making the changes needed so as to present food microbiological analysis as a more attractive investment area.

Some of the problems are described in the next few pages. There is no attempt to assign relative importances. In any case, these will vary according to the nature of any instrument being considered. If the reading seems depressing, keep in mind that I

---

* Based on estimates for the USA in *ISA Transducer Compendium*.[1]

† Based on listings in 1970 and 1972 Bureau of the Census, and 1975 Statistics Canada publications.

am trying to show the unhealthiness of our present situation and the dimness of any hopes for significant advances in the near future. Food microbiology now encompasses so many interests, so many different objectives and techniques, that in our present hopes for wholesale instrumentation we are trying to push it up a gradient altogether too severe for its sprawling frame.

THE CAPABILITY PROBLEM

Promises of increased power or capability (for example, sensitivity or accuracy) have always been major stimuli for the development of instruments. Indeed, the continued ability of new instruments to extend scientific horizons is vital to the existence of research scientists and scientific research. While not everyone at the bench is a research scientist, particularly in microbiology, the unexpected phenomena uncovered by a new tool are usually quickly shown to be relevant to health, production, control, warfare, etc. When this happens, many laboratories find they need the instrument, and the market grows to a very significant size. Entrepreneurial instrument companies have always operated with this in mind and have invested heavily in research and development (R&D) on instruments that extended existing capabilities rather than tried to fill the apparent needs of the moment. To those companies that were also astute enough to be successful have gone the greatest rewards for their investments. To these also we largely owe our standard of living, for it is difficult to imagine industrial or medical productivity rolling along as it does without in-line spectrophotometers, automatic analyzers, computers, etc.

Such companies win the jam. Many other survive, albeit quite nicely and with greater security, on the bread and butter of existing technology. At this level, development investment is minimal, and the market and probable returns on capital are more predictable. Only a promise of very high returns makes a stable and well-founded company speculate in a new area of technology, and it is almost axiomatic that the big market from which this high return comes only develops if one provides a previously nonexistent capability.

In sciences where there is now heavy instrumentation, the development of instruments has always been first to extend human capabilities beyond those with which we evolved and then beyond those we can control or even properly comprehend in real time. Whereas, unaided and under the right conditions a human can

feel thicknesses of 15 to 20μm,

see particles 20μm in diameter,

perceive grey changes of about 1 percent,

taste H$^+$ ion changes of 2.4 $\times$ 10^{-6} molar, e.g. the change from pH 7.0 to 5.6,

hear sounds ranging over about 16 kHz, and level changes down to 3 db,

count one or two scintillations per second from energetic particles,

pour liquids one at a time without getting wet,

add 1987 to 2374 in about 4 seconds,

investment in instrument technology has provided us with interferometers, the microscopes, counters and pulse analyzers, autoanalyzers, IBM 360s, and so on. In many cases, initial concepts may not have been commercially oriented, but ultimately it is to commercial companies and their recognition of the market potential of increased analytical capability that we owe their reality.

This brings us to quantitative food microbiology and one of our major problems, for there are good reasons for doubting that instrumentation will lead to any improvement in limits of detection, sensitivity, or accuracy* in the sense that we interpret these terms at present. Any instrument companies tentatively probing into this field, armed with their experience at detecting and analyzing minute quantities of materials, have surely been quickly dismayed by the demands of food microbiologists. What other branch of science might demand, for example, a level of detection equivalent to one *Salmonella* cell in 25 g of meat? Here, the total cellular mass fraction of *Salmonella* may be only 2 $\times$ 10^{-14}, and most of the remaining 0.999,999,999,999,98 parts are physically, chemically, and biochemically very similar to *Salmonella* cells. The mass fraction of specific material (an enzyme,

---

* The concept of limit of detection in an analytical procedure is covered in more detail in Chapter 3.

perhaps) on which an analysis for *Salmonella* cells might be based will be even lower — probably less than 10^{-16}. Commercial instrument engineers will quickly decide that there are many fields with better signal to noise ratios more suited to their capabilities. Detection at this level is certainly an extreme requirement, feasible in only a few areas of science. They will note, however, that the food microbiologist can laugh at their discomposure, for he has no difficulty at all in detecting one *Salmonella* cell in 25 g of meat, using just his fingers, his eyes, and a few dollars worth of materials. Without any investment in costly instruments, his analytical capability is already greater than can be offered by almost any other branch of science. All that seems to be left to instrument manufacturers is mechanization of existing techniques and, though this may sound simple and straightforward, there are pitfalls as will be shown in the next subsection (The Engineering Problem).

What other science would also demand to know the number of cell clusters of *Staphylococcus aureus* in a frozen airline dinner rather than their total relevant metabolic activity, i.e. the amount of toxin they can produce, in order to determine whether or not that dinner is hazardous? Commercial instrument engineers will quickly decide that microbiologists must be convinced of the property of *Staphylococcus aureus*ness through the application of a battery of tests — none of which actually represent someone being ill. The existing methodology (plating and incubation on Baird-Parker agar, for example, followed by subculturing, further incubation, and coagulase tests) is impossible to simulate mechanically in instruments of realistic price. Image-analyzing computers, for example, are already valuable in many areas of biology, but much more expensive development is needed before their performance is comparable to that of a human eye roving amidst bits of gristle, pea shells, and fish scales, looking for a few outgrowths of particular shape, color, and marginal appearance. The microbiologist's simple manipulations of sterile pipets, bottles of diluent, and growth media also are most difficult things to simulate mechanically. Manufacturers deciding that realistic detection systems must be based on other properties of *S. aureus* soon find that while they could produce instruments capable of satisfactorily detecting metabolic products of this organism such as coagulase or deoxyribonuclease, at levels corresponding to, say, 10^6 cells/g, detecting them at regulatory levels of 100 cells/g or less is quite another matter.

The problem is compounded by the certainty that the performance of any instrument will be assessed on the basis of its ability to provide data correlating with conventional plate counts. In Chapter 4, however, it will be shown that such correlations (if they exist at all) are likely to be accidental exceptions rather than the rule.

Commercial investigations into the field must certainly show that as long as food microbiologists remain committed to their present data concepts, no practical instruments can be developed that will provide the levels of sensitivity, selectivity, and confidence they apparently now enjoy, much less exceed them, unless microbiologists develop the ability to pay astronomical prices for those instruments.

Being unable to provide customers with something better than they already have, or to manufacture a need by extending horizons, instrument companies see little point in wasting good money in the area. There is no shortage of other areas in which to invest development money, and it is in favor of these that manufacturers continually shrug their shoulders at food microbiology. This is how things are likely to remain for as long as we continue to assume that only our existing ways of assessing food quality provide us with the data we think we (and the consumer) need.

THE ENGINEERING PROBLEM

An instrument engineering group looking at food microbiology for the first time will quickly arrive at a decision point. They will observe that the quality of food appears to be assessed on the basis of enumeration of particular groups of viable organisms and that the data are generally compared with statutory limits on numbers or used as predictors of keepability, saleability, etc. They will observe that two major avenues appear to be open to development. On the one hand, they may develop instruments that merely carry out the existing analysis in more or less unaltered form but faster, more accurately, more reliably, less expensively, etc. On the other hand, they may develop instruments based on the biochemical, immunological, radiometric, spectrographic, chromatographic, and other techniques that have proven so successful in other areas of science. After producing an instrument with adequate capability, they must then ensure enough data can be provided that food microbiologists are convinced of its acceptability. They will see that to be acceptable, data from a new instrument must be interpretable in terms of the existing enumerative methods so that it may substitute them right away. Ideally, it must then be seen to be more powerful so that it can promise conceptual advances over the conventional procedures.

The first avenue appears to be mainly an engineering problem: make the machine work sweetly, show that it does just what a human does, only better, make it affordable, and — like the proverbial mousetrap — the scientific world should beat a path to the sales door. The second appears to be mainly a costly scientific problem. The basic instruments or techniques probably already exist but need to be applied to food microbiology in a research program that will determine the methodology, the necessary accessories, the limitations, the correlations, and so on. Both avenues have been trodden by various companies at different times.

If the former avenue is chosen, it soon becomes apparent that there are going to be severe engineering problems in mechanizing at a realistic cost any series of operations remotely resembling conventional quantitative microbiological procedures. The operations carried out by a food microbiology technician are deceptively simple. They are also deceptively few. Let us look at a basic analysis from an operational point of view.

Almost all of the popular microbiological analyses are variations on the following basic set of operations:

1. Quantify the sample (weight or volume).
2. Transfer microorganisms from the sample to growth medium.
3. Prepare a range of dilutions.
4. Incubate, i.e. take off-stream for a while.
5. Detect and enumerate regions altered by microbial growth.
6. Subculture and carry out other tests.

Many analytical procedures exist under the names of plate counts, most probable number (MPN), droplet, agar sausage, and enrichment techniques, in endless variations, all however, differing from one another only in the extent that they encompass or avoid the implications of each of these operations. Methods for use in the field generally tend to avoid as many operations as possible. At the other extreme, official methods tend to give full consideration to each stage.

These few descriptive operations are only very broad generalizations, however, and when we start to break down an analysis into the individual operations carried out by a technician or working machine, we quickly reach a disconcerting level of complexity. Suppose we were foolish enough to construct an instrument duplicating most of the operations presently carried out by a human technician during a simple standard plate count (SPC). To complete the automation, the instrument should also prepare its own media. If we classify individual operations as being at the level of, say, "Pick up pipet," or "Remove lid from Petri dish," we can break down the analysis into the list shown in Table 2-I.

Without adding any operations to describe the possible picking out of colonies for subculturing and further biochemical tests, Table 2-I contains 974 individual instructions. Technicians, of course, generally anticipate the final result so that they may reduce the number of dishes, dilutions, and total number of operations they carry out. However, if our instrument is to assist in a normal laboratory, it will eventually meet all possible sample contamination levels and must, therefore, be able to pursue this full scheme of operations.

Such an instrument, if it worked and could be purchased at a realistic price, would presumably be accepted easily in every laboratory. Because it merely duplicates an existing procedure, few suspicions regarding the possible production of spurious data could be raised against it. The only evaluations each laboratory might wish to make would be, for example, that it ran smoothly without breakages, spills, etc.; it did not become horribly contaminated; its counter matched (!) the human eye; its cost/benefit comparison was favorable.

The cost of developing and manufacturing this instrument

Table 2-I: APPROXIMATE LIST OF THE OPERATIONS INVOLVED IN MAKING A STANDARD PLATE COUNT.

AGAR MEDIUM PREPARATION
Bottle to washer.
Wash. Dry.
Bottle to prep. area.
Dry medium to prep. area.
Open medium package.
Measure out medium.
Add medium to bottle.
Stir. Pick up closure.
Close bottle.
Bottle to sterilizer.
Sterilize. Cool.
Bottle to work area.

DILUENT PREPARATION
Bottles to sterilizer.
Sterilize. Cool. Wash.
Large bottle to prep. area.
Dry medium to prep. area.
Open medium package.
Measure out medium.
Add medium to bottle.
Stir. Heat.
Insert pump.

> Diluent bottle to prep. area.
> Centre pump outlet.
> Pump diluent.
> Pick up closure.   × 9
> Close bottle.
> Move to sterilizer.

Sterilize. Cool.
Move to work area.

PIPETS
Dirty pipets to sterilizer.
Sterilize. Cool.
Move to cleaning area.

> Pick up pipet.
> Place in blower.
> Blow out plug.
> Place in washer.
> Move to dryer. Dry.   × 8
> Move to plugger.
> Pick up pipet.
> Place in plugger.
> Plug. Place in can.

Table 2-I   *(continued)*

PIPETS *(continued)*
Can to sterilizer.
Sterilize. Cool.
Move to work area.

PETRI DISHES
Open carton.
Remove bag of dishes.
Slit bag. Dispose bag.
Dishes to work area.

BLENDERS
Dirty blender to sterilizer.
Sterilize. Cool.
Blender to cleaning area.
Remove closure. Empty.
Blender to washer. Wash.
Replace closure.
Blender to sterilizer.
Sterilize. Cool.
Blender to work area.

SAMPLE
Receive sample.
Record sample identification.
Sample to work area.
Open sample container.
Blender on balance.
Remove blender closure. Weigh.
Transfer sample to blender.
Weigh. Compare with required.
Add/remove sample. Reweigh.
Compare with required.
Add/remove sample. Reweigh.
Remove diluent closure.
Add diluent to blender.
Close blender.
Blender to blender base.
·Blender on. Time. Off.
Blender to work area.
Diluent bottle to cleaning.
Dispose sample container.

Table 2-I. *(continued)*

DILUTION, PLATING

Remove blender top.

Pick up pipet.
Connect pump to pipet.

Place pipet in liquid.
Withdraw volume.
Compare with required.
Add/remove. Compare.
Dispense Petri dish.   × 2
Label dish. Remove lid.
Transfer pipet to dish.
Release liquid.

Replace pipet in mixture.
Withdraw volume.   × 8
Compare with required.
Add/remove. Compare.
Open diluent bottle.
Place pipet over bottle.
Release liquid.
Close bottle. Mix.
Disconnect pump.
Pipet to "dirty" can.
Remove bottle top.

Open agar bottle.

Pour agar.
Close Petri dish. Mix.   × 16

Close agar bottle.
Cool dishes.

INCUBATION

Stack dishes. Invert.
Incubate. Time.
Remove dishes to work area.

COUNTING, RECORDING

Select plates (16) for one sample.
Illumination on.

Dispense plate.
Estimate count.
Reject if $30 > $ count $ > 300$.   × 14
Dispose plate.

Table 2-I. *(continued)*

COUNTING,   RECORDING *(continued)*

Hold plate if 30 < count < 300.
Pick up marker.
Scan plate and count colonies.
| Mark colonies. | Average   × 165

Put down marker.   × 2
Pick up pen.
Read dish label.
Record dish count and dilution No.
Calculate count.
Record count.

Record average.

CLEAN UP
Dirty pipets to prep. room.
Dirty bottles to prep. room.
Dirty blender to prep. room.
Agar bottle to prep. room.
Petri dishes to prep. room.
Sterilize dishes.   Dispose.

would be enormous, of course. Commercially, it would never be seriously considered, for the overall process can be simplified and streamlined with advantage. Operations may, for example, be eliminated entirely from the instrument through the use of commercially available ready-mixed media or ready-poured plates. The technician can be asked to weigh samples, feed materials into the instrument, count plates, record data, and so on. Such simplifications affect only the instrument, not the overall process, since those parts of the burden are merely shifted into the factory or to the technician.

The engineering group is now faced with a trade-off. By reducing the number of operations the instrument carries out its complexity and, therefore, its development and manufacturing costs will be reduced. This should make instruments more sellable since they will probably be so expensive and so marginally affordable by microbiologists that the selling price will be determined more by manufacturing costs than by what the market will tolerate. As the selling price decreases, more laboratories will be attracted to it and sales should increase. It may then be commercially attractive to shift as much as possible of the process into the factory by developing software (consumables) specifically for the instrument. The instrument itself (hardware) may then be even further simplified and its price brought further down. At the same time, it becomes capable of using only the company's software, from which continuing profit will be derived. This is much more attractive than the one-shot profit otherwise obtained from sale of the hardware itself. In the limit, the company could decide to rent out or even give away instruments on the basis of expected profits from the consumables sales they would generate.

Against these pleasant commercial extrapolations, however, rise two other factors. First, as the instrument's complexity decreases, so will its capability and versatility, its value to most laboratories, and its sales potential. Second, the more the process is streamlined in favor of engineering feasibility, the more it must deviate from the process currently carried out by technicians. This will, unfortunately, lead to much closer scrutiny of the relation between its data and those from the existing method. In other words, the closer the instrument is brought to engineering practicality, the greater will be the eventual cost of validating its performance in the light of accepted practices.

The trade must be made on these competing factors. Most food microbiology laboratories will claim that purchase of an expensive instrument can only be justified if it takes over a large proportion of the laboratory workload. For an average laboratory, this could mean expecting the instrument to be capable of working with samples containing up to 10^9 organisms/g and of analyzing samples for perhaps four different parameters, e.g. SPC, coliforms, *Staphylococcus aureus,* molds. The specifications for an instrument of even modest capability are bound to include, at the least, programming, indexing, transporting, sorting, fluid dispensing, agitating, incubating, detecting, and means for recording. All of these must be designed to work aseptically in the presence of contaminated sample materials, continue operating reliably with lumpy suspensions, and not require two days to set into operation.

The engineering requirements are too severe for commercial

practicality. If monies were no object, such instruments could be in every laboratory. But the engineering cost is too great for the market to support. Approaches to practicality have been made, but it would seem that as time goes on, developing instruments that simply replace technicians in conventional enumerative analyses will be a less and less attractive avenue for R&D interest.

The trade-off has, in fact, always materialized on the side of simplicity, minimal cost, and attack on local problems identifiable among the whole. Thus, we have portable media preparation units, Petri dish filling machines, Stomacher® blenders, disposable Petri dishes and pipets, plate streaking/diluting machines, electronic colony counters, and data handling terminals. All of these developments have been useful advances. None of them, however, eliminate the need for a technician at more than one portion of the process at a time. Nowhere are they integrated into one comprehensive analyzer. None of them attack the basic problem of food microbiology.

THE CREDIBILITY PROBLEM

The apparent alternative to machines based on existing food microbiological procedures is development of instruments using conventional analytical techniques such as spectrophotometry, chromatography, or radiometry, capable of yielding immediate data about microorganisms at interesting concentrations in foods. If such techniques exist, they have become neither popular nor commercially developed to the point where they are seen to be really viable alternatives to Petri dishes and broths. This, hopefully, will not always be so, and there are already limited areas where nonmultiplicative techniques are used successfully by suitably motivated laboratories. A broader discussion of the problems of microbial detection by physical or chemical methods will be found in Chapter 3. It is realistic to expect that methodology developments (for example, improved extraction procedures) could eventually permit several analytical techniques to yield data at detection levels relevant to, say, regulatory limits of microorganisms in foods.

Development of such techniques would seem to be a most natural avenue for an instrument manufacturer since a large part of the final instrument may already be in production as a company line, and its technology will be well established. However, as has already been mentioned, this approach merely trades instrument development costs for the funding of a methodology development and scientific validation program. The technique is now so far removed from the conventional process that it will receive the most critical scrutiny and suspicions about the reliability of its data.

There are limited areas where the cost/benefit of a physical technique may be so favorable and the accountability so low that it can be readily taken up by a laboratory. The information given by a rapid technique for detecting incipient putrefaction in raw meat, for example, may be considered better than no information at all when rapid decisions must be made regarding its use in the different product lines within a factory. In general, however, the credibility of a new technique may be extremely difficult, if not impossible, to establish in the context of food microbiology as it exists today.

Chapter 4 emphasizes that alternative methods of analysis are incapable of yielding data correlating with the plate counts (or equivalent methods) on which our present concepts of food quality and standards are based. The problem lies, however, not with many of these methods but in the inadequacy of the count itself as a descriptor of quality. It is a unique type of analysis with no parallels in the physical world, and its datum has a unique and illusory significance that is far narrower than we are generally misled into reading. It is most unfortunate, then, how we have been so conditioned by concepts and doctrines handed down throughout the history of food microbiology that we instinctively rush to check the performance of an alternative analytical method against the plate count and regard it as inferior and unacceptable when no correlation materializes.

The would-be developer of instrumented methods faces this credibility gap if his methods are based on conventional physical or chemical processes. The techniques an instrument company can provide will generally not correlate with plate counts to the extent that food microbiologists are willing to accept them. A research group may produce beautiful curves from which the plate count of a pure culture may be calculated with confidence, but a quality control microbiologist in a food factory who bor-

rows the instrument tries it out on TV dinners and barbecued
chicken and hands it back with tactful criticism. He does not get
the beautiful lines. He cannot interpret the results in the light of
his existing experience because the correlations are so poor. He
is not convinced he can effectively control the quality of his
factory's products unless he accumulates large quantities of
parallel data for several months; in all probability, he does not
have the facilities to take on this extra burden. And he knows
that because regulatory limits are based on plate counts, if the
situation arose, a prosecution lawyer could turn his efforts at
microbiological analysis into an object of ridicule.

This credibility problem is probably the most important obstacle to the instrumentation of food microbiology. Wider acceptance of physical, chemical, biochemical, or immunological
methods is essential if throughputs and analysis times are ever to
be significantly improved. However, no self-respecting microbiologist can afford to drop his present techniques which, in
this immediate context, yield precise information, in favor of
others yielding only probabilities. A plate count datum of 3.10^5
coliforms/g, for example, will be considered more precise and
will certainly be more acceptable than an alternative one based
on glutamate decarboxylase activity that at best may only indicate a 70 percent probability of the count being 3.10^5.

In reality, a metabolic activity datum may be rather more
pertinent to potential spoilage or hazard from a product than a
count of organisms. But it loses much of its credibility when we
try to interpret it as a microbial count because (for reasons
discussed in more detail in Chapter 4) any apparent correlations
between different types of microbiological data are quite fortuitous. It certainly is the most unfortunate problem, since it
exists only because of the persuasiveness this image of the count
holds for us as a food quality descriptor. It so prejudices our view
of food microbiology that we find more valuable concepts wanting because we distort them until they ungracefully fit the numerical screen. It is often difficult to accept that data not correlating with counts usually fail to do so because they are actually
more accurate or relevant. This lack of credibility is a powerful
deterrent to the acceptance and therefore to the further de-

velopment of commercial instrumented methods in food microbiology.

THE INCENTIVE PROBLEM

New products are most attractive when they realize maximum benefit from the company's existing know-how, tooling, and organization. Putting as many components as possible from existing company products into a new line increases the efficiency of the production department. It minimizes development time, investment in tooling, and the number of unexpected problems occurring during development. It smooths such important peripheral aspects as stock control and the training of service personnel. Moreover, the more naturally a new product fits into the company's existing area of operations, the more reliable the predictions are that its sales organization can make about the potential market and its readiness for the change.

It is unfortunate that machines or instruments based on conventional quantitative microbiological processes will be quite dissimilar to existing lines in most scientific instrument companies. Developing instruments for food microbiology, therefore, will generally be costly and more speculative than, say, updating a spectrophotometer line by the introduction of a new lamp, or a gas chromatograph line by the introduction of a new programmable oven. Most companies thus prefer to take this latter approach, to the misfortune of food microbiologists.

The patent literature is littered with descriptions of devices for preparing, pumping, and mixing microbiological media; making dilutions, multicompartment growth vessels, enumerating colonies, etc., and various combinations of these. Notwithstanding the probability of devising some ingenious solutions, it is unlikely that any really new technology is needed to develop instruments for conventional microbiology. This is a pronounced disadvantage because although great expense may be incurred developing and accumulating know-how for a practical working instrument, there will remain a significant risk that the product may be found to infringe other patents or that any patents obtained by the company for its own product will only be of discouragement value and will be found to be insubstantial if litigation later occurs.

Faced with heavy R&D costs in an area of questionable pro-
tectability, only the promise of rapid returns from a lucrative
market is likely to provide sufficient incentive for further commercial interest. The majority of food microbiology laboratories,
however, have a very poor financial status. Many are perceived
to exist, for example, not because their existence tangibly increases company profit but because of some far more nebulous
calamity they may help avoid. Their most obvious activities in the
company are generally negative, and they have a poor bargain-
ing position with senior management. They are manifestly unable to rush into the purchase of expensive pieces of equipment.
Commercial R&D in microbiology is, therefore, biased heavily in
favor of the much more rewarding clinical area where financial
statures are higher, and the problems and requirements are
better defined and generally more soluble from an engineering
point of view.

THE VERIFICATION PROBLEM

Research, development, validation, publicity, and reeducation
programs of sufficient scope, intensity, and persistence to
reorientate food microbiologists, food manufacturers, regulatory agencies, consumer associations, etc. to the acceptance of
instrumented methods and new forms of data will be so costly as
to be capable of being carried through effectively only by the
largest of instrument companies. Federal and other governmental research grants to industry notwithstanding, the
R&D budgets from which these operations are financed commonly run at up to 10 percent of sales in large companies;
significant proportions are invested in long-term projects of the
kind required in food microbiology. However, large companies
grow to the size and capability they enjoy only through the
astuteness of their business management and, as we have seen,
such businessmen are unlikely to pay food microbiology much
further attention once they become aware of the technical complexities, the conservatism, and the poor financial status of the
field. Large companies go after large markets, and there are
more receptive and more attractive areas of science.

Notwithstanding, smaller companies are capable of producing
ingenious solutions to many everyday microbiological problems

and would be quite happy to enter the food microbiology instrument market. Lacking the financing power of larger companies, however, they require an immediate market. But a market cannot materialize immediately, owing to the reserve shown by the mass of food microbiologists in the face of lack of adequate corroborative data and acceptance by the regulatory agencies. The solution appears to be straightforward; the regulatory agencies must be convinced into accepting the technique. To have an instrumented technique written into official or even recommended methods would at once open the door to the widest possible sales field.

This is not easily done. Regulatory agencies cannot change standards or recommendations as soon as somebody comes up with a bright idea. Their responsibility to the public demands that changes in statutory requirements can be made only when they possess the fullest possible knowledge of all pertinent factors. Thus, it seems almost inevitable that changes will only be made after careful (slow) deliberation and only in the light of favorable results from collaborative studies made in a number of laboratories. The burden presented to an instrument company is that of financing the loan of expensive prototype instruments for many months while these collaborative studies are made. This is done with the knowledge that few sales are likely to be made until a report of the study has been written and published, and microbiologists are at least aware that their regulatory agency is warming to the idea of changing its requirements. Small companies, operating more on a monthly or even weekly cash flow basis, simply cannot afford this delay before their development investments begin to generate sales.

ALTOGETHER

Altogether, the prognosis is not good. There is no doubt that the future efficiency of food microbiology will be strongly linked to the achievable level of instrumentation and therefore to the level of commercial interest that can be generated in the subject. As things stand, however, we have a vicious circle. The market potential is poor because the advantages offered by achievable automation are marginal. The R&D budgets needed to produce acceptable instruments at realistic costs are enormous. Only large companies have the resources to develop satisfactory engineering solutions or to finance validation programs for more easily engineered solutions. Large companies are not interested because the market potential is poor . . . and so on. The circle could block significant advances in food microbiology instrumentation forever.

What we get are gadgets, ingenious solutions to parts of the problems of food microbiology, devices operating sufficiently close to conventional processes that they are accepted without too much difficulty and produced by small companies that manage to find enough ready market to pay their way.

I believe the slow infiltration of such devices does food microbiology no good. Even if these gadgets manage to be successful enough to ease laboratory burdens a little, they are serving only to entrench the framework of methods and concepts on which food microbiology is currently based — a framework which, itself, is the principal inhibitor of progress in automation. If we are ever to have in our laboratories instruments of sufficient capability and reliability that we can entrust them with guarding the quality of our foods, we must try to rearrange this framework so that holistic engineering solutions to problems of microbiological automation are feasible.

It will be argued that to consider changing a conceptual framework merely to accommodate a desire for automation is absurd unless it can also be shown that the new framework is at least as capable of controlling food quality as the old. Just because our existing concepts date from the nineteenth century does not necessarily mean that they are outmoded at this stage of the twentieth century or that we should cast them off before the twenty-first. It is not enough that the clutter of methods embodied in them presents us with the most appaling problems of instrumentation. However, I believe the concept towards which this book is directed *might* be capable of safeguarding the public more effectively from the effects of foodborne microbial contamination; it happens that the methods derived from it are also fundamentally more suited to instrumentation than traditional microbiological analyses. In Chapter 4, it will be shown that while traditional methods serve to control food quality on a broad statistical basis, they provide us with, at best, only one-fifth of the information we need in order to be able to describe the individual quality of a food. The approach I shall describe would provide the same statistical protection; in addition, it can provide *all* of the information we need to measure the quality of foods in individual situations.

There is no doubt that any increases in the level of instrumentation that allow food microbiologists to analyze more samples per day will improve their ability to monitor raw materials, plants, and products and obtain a better overall picture of the distribution of contamination, etc. To this extent, further development of piecemeal instrumentation will improve the efficiency of food microbiology. But should anyone be allowed to persist in painfully attempting to mechanize or automate traditional microbiological methods if, in so doing, food microbiology is pushed farther and farther away from consideration of more efficient alternative approaches to the subject? If there is any possibility that another approach could reduce the amount of good food destroyed or diverted unnecessarily, reduce incidences of food poisoning by providing protection on an individual rather than statistical basis and yet facilitate the development of practical instruments, then we should halt even the small amount of effort presently being devoted to piecemeal engineering solutions to microbiological problems.

The effort required to alter the established direction of food microbiology would be enormous. The longer we delay, and the more entrenched we become through the development of these piecemeal solutions, however, the more difficult such a reorientation will become. It is imperative that all concerned with the microbiological quality of food examine whether our present objectives, methods, and achievements are adequate for the coming century and whether we should even consider automating food microbiology as it stands at the moment.

TROUBLE WITH AGAR

The history of science shows repeatedly that concepts that work well when they are first put forward become increasingly inadequate the more facts we try to fit into their frameworks.

Much of the concept and methodology of food microbiology was formed at a time when metals were detected by group separation; radioactivity was lying around in a drawer waiting to darken some photographic plates; and the aerospace industry fluttered in the breeze at the end of a waxed string. In this age of atomic absorption spectrometers, scintillation counters and 747s, we should not overload food microbiology by expecting it to compete on modern technological terms unless its structure has been proven capable. I don't think it has or will be, and while not wishing to be too irreverent, I would suggest that a principal source of weakness in food microbiology has been brought on by reliance on the structural properties of agar. It is a magnificent material, to be sure, but its almost ubiquitous use as a foundation for the science may have led to the construction of a rather shaky edifice.

In the beginning, the methods used for microbiological testing of foods were determined by the necessity of working by hand. At the same time, the small concentrations of detectable materials presented by most microbial cultures made their detection by available physical or chemical methods difficult, if not impossible. These two factors favored the development of the enumerative (plate, most probable number, presence after enrichment, etc.) methods. In these, detection problems are minimized because the microorganisms function as their own amplifier (Chapter 3), and the actual detection (evaluation) step can be made easily, using just the human eye. There is no doubt that agar based methods are much preferred by microbiologists whenever it is possible to use them; agar lends a tangibility to microorganisms that other methods of visualizing them do not.

To actually "see" microorganisms as colonies on an agar plate or as dendrites in a tube, for example, or to be able to distinguish many of them and to be able to count them off this way, like little individuals, led to a great feeling of closeness to them. The rapport thus established between microbiologists and microorganisms lent enumerative data a remarkable aura of credibility. Once microorganisms were established as "the enemy" in food, it was an immediate and instinctive step to assess any danger on the basis of their number. The subconscious impact of a count seems to go far beyond the value of the information it actually carries, however. While it has always been regarded by many microbiologists as just a useful item of information, to others (and certainly to laymen), it is now the essence of food quality. Journalists are well aware of the direct link between numbers of microorganisms and human emotional responses, and the very tendency for food standards to be set on a numerical basis emphasizes its value to the average person.

Nevertheless, there has always remained the difficult problem of assessing the true significance of enumerative data to human health or to organoleptic acceptability. Perhaps equally important today, however, is the fact that the entrenchment of enumerative methods has created a powerful barrier to the development of mechanized or automated methods in food microbiology. In this sense, the suggestion of Fannie Eilshemius Hesse[2] regarding the usefulness of agar jellies for the growth of bacteria, which has been looked on as the miracle of microbiology, might be viewed in future as being one of the gravest disservices rendered to the science. Agar made things too easy for microbiology; the methods that were developed following its availability were too perfectly suited to the capabilities of human hands, and the data it yielded were too suited to the visualization preferences of the human mind. Our dependence on the enumerative data concepts that resulted now stands in the path of change. Without agar, microbiology might naturally have developed methods that are fundamentally more easily mechanized or else its techniques might have remained so tedious to carry out that novel instrumented methods would have found a climate of opinion much more enthusiastic for changes.

The true tragedy of agar is that it stimulated the development of a whole science around a unique type of analysis — the *plate count.** Here was an analytical technique of seemingly utter simplicity, almost completely detached from expensive hardware requirements, with an exquisite sensitivity, even today almost unparalleled in the rest of the sciences (Chapter 3). A

---

* Of course, there are many microbial counting methods. I simply use the phrase *plate count* to cover all methods using solid growth media.

technique which, unfortunately, from its very nature, provides data so unique that they can be related to no other analytical data than those from other plate counts.

The celebrated Robert Koch seems to have been quite content to absorb all the honors that, by right, were due at the time to Fannie and her husband Walther.[3] To him also, therefore, might be ascribed the unbecoming credit of leading the way to an analytical backwater from which we shall only escape with prodigious effort.

REFERENCES

1. Minnar, E. J. (Ed.): *ISA Transducer Compendium.* Plenum Pr, New York, 1972.
2. Hichens, A. P. and Leikund, M. C.: The introduction of agar agar into bacteriology. *J Bacteriol 37:*485, 1939.
3. Koch, R.: Die Atiologie der Tuberkulose. *Berl Klin Woch 19:*221, 1882.

Chapter 3

ON THE FEASIBILITY OF INSTRUMENTING FOOD MICROBIOLOGICAL ANALYSIS

THE SATISFACTORY quality control of food depends very much on the ability of both manufacturers and regulatory agencies to measure its physical, chemical, biochemical, and microbiological properties. For the first three types of properties, many instruments have been developed that allow the analyses to be carried out either directly or within minutes of obtaining a specimen. Many of these have also been completely automated to the extent that they are used in-line for the control of manufacturing processes. In various branches of the food industry alone, control of weight, rheology, viscosity, temperature, specific gravity, crispness, firmness, tenderness, clarity, color, refractive index, gelling time, moisture, solids, ash, pH, total acid; chlorine, salt and other minerals; fats, oils, unsaturated fats, fiber, carbohydrates, protein, the vitamins; metals such as copper, iron, lead, mercury and nickel; many flavor or odor modifying substances such as acetoin, alcohols, aldehydes and diacetyl; enzymes such as catalases, diastases, invertases, lipases, pectin esterases, peroxidases; a variety of acaricides, antibiotics, bactericides, disinfectants, fumigants, fungicides, herbicides, insecticides, PBBs, PCBs, and many other significant parameters are carried out routinely, either using in-line instruments or in quality control laboratories during the manufacturing period. There are, however, few instances where any kind of microbiological control is feasible during manufacture. Apart from exceptions such as the use of sensitive reduction indicators like resazurin to *estimate** the quality of meat and milk products, microbial control during manufacture is effectively confined to

---

* The word is italicized to emphasize that, particularly in the meat industry, such tests are used as substitutes for plate counts rather than in their own right. There is, in fact, no cause for supposing that when resazurin data do not correlate with plate count data they are any less valuable as indicators of the ongoing microbial activity.

43

slow processes, which themselves often depend on microbial growth, e.g. beer, cheese, yoghurt.

Unlike many of the other parameters mentioned, the levels of microorganisms in foods do not normally remain constant with time. This difference is fundamental to the problem of setting national or local quality standards, to the instrumentation of food microbiology, and to the general relevance of food microbiology in the physiological impact of foods on the consumer. The toxicologist interested in mercury contamination, for example, can be fairly sure that the level of total mercury in fish at the instant of consumption is the same level that is in the fish when it passes through the packaging plant. Statutory limits for such nonmicrobial contaminants are arrived at simply by taking some fraction, e.g. 1 percent, of the lowest level suspected of causing a response in humans. Analytical methods developed for such contaminants, therefore, need only have limits of detection corresponding to (statutory) human tolerance levels, and the analytical datum has immediate relevance to the consumer regardless of the instant at which it is obtained.

The same is not true for microbial contaminants. The consumer who becomes ill from staphylococcal toxins may need to ingest 10^9 or 10^{10} cells of *Staphylococcus aureus* (or rather, the toxins produced by, say, 100 g of ham containing 10^7 or 10^8 cells/g). However, in order to achieve an acceptably small percentage of incidences of poisoning across the country, a statutory limit some 10^6 times lower than this may be required so as to accommodate the tendency of *S. aureus* to proliferate in that food. Several thousands or even millions of *Salmonella* cells may need to be ingested before most persons would contract salmonellosis; however, the statutory level for *Salmonella* in milk powder must be very low, e.g. absent in 100 g, to accommodate the ease with which this organism flourishes in foods made up from milk powder. Similar examples exist for almost all forms of foodborne microbiological contamination. The food microbiologist certainly faces a difficult problem. To use microbiological data this way implies an ability to extrapolate from the instantaneous value at the moment of sampling to a future value a million times or so greater. But even the slightest deviation from an accepted model of microbial growth can make such extrapolations worthless; the range is just too great.

Notwithstanding, if food is to be tested at all for levels of microbial contamination, the analyses must be made somewhere and at some instant. From the consumer's point of view, it might be preferable to test all food immediately before consumption, either by means of inexpensive household instruments or by including indicators in the actual retail packages. One can imagine packages, for instance, bearing labels like, "Do not eat these sardines if *Botulinum* strip shows red." Unfortunately, food manufacturers would strongly resist the incorporation of such negative imagery into their products, and the potential for controlling consumption of contaminated food this way is probably very small. A more practical control focuses on the raw materials and the ability of the processing line to destroy intrinsic contamination and prevent more from entering the product. In an ideal world, universally strict adherence to good manufacturing, distribution, and storage practices would virtually eliminate the need for microbiological testing. The potential for achieving this utopian perfection is also small. Microbiological analysis of raw materials and finished products will continue to be necessary for as long as humans continue to be involved in the production and handling of food or wherever there is no administrative control over imported materials. Only microbiological analyses can provide the necessary information for decisions on acceptance, corrective actions, distribution, and so on.

Unfortunately, as long as analyses are not made at the instant before consumption, the capricious growth characteristics of microorganisms and their products must be taken into account. Statutory limits for microbial contaminants thus continue to be set at very low levels, compared with nonmicrobial ones, to allow for the probable growth of physiologically active parameters through many orders of magnitude during the time between analysis of a food and its consumption.

Detection of microorganisms at these very low levels is quite possible. The traditional microbiological methods, which are in daily use in thousands of laboratories, have this capability. But the analyses take time, for their success depends on exploitation of the leisurely self-amplifying property of cells. They are quite unsuited to in-line or direct control of manufacturing processes measured in minutes or hours. Thus, traditional microbiological methods tend to be used in factories as they are in regulatory work, mainly as statistical indicators of the general standard of processing. The products to which they relate have often been distributed and consumed before the microbiological data are available; at other times, stock must be held in expensive cold storage until the data are available. There is, in consequence, much recognition of the need for faster methods of microbiological analysis and particularly of greater throughputs to allow satisfactory coverage of products. Most microbiologists look to instrumentation as the eventual means of satisfying both these requirements.

Just how well founded, however, is the hope that instrumentation will one day provide the speed and sensitivity needed for real-time control of product quality? The idea that the outlook is not too good has already been introduced in Chapter 2 under The Capability Problem, but we should attempt to define the size of the problem, if only to help ourselves understand some of the commercial instrument manufacturers' reluctance to invest heavily in food microbiology instrumentation.

In the next few pages, I shall try to examine the sensitivity/speed trade-off facing would-be developers and put the detection requirements of microbiologists in some perspective. Much of the argument is verbal and hopefully easily readable. A little math was unavoidable, however, and seemed to fit into the running text much better than as an Appendix. To readers who dislike mathematics, I would suggest at least reading along with the text in order to become familiar with a few terms that are used in later chapters.

LIMITS OF DETECTION

There is a great tendency among microbiologists to look at very "sensitive" techniques used in some other branches of science and ask why it is that these techniques have not been used as bases for microbiological analytical methods. When so many detectors of radioactive disintegrations, photons, chlorinated compounds, and so on can be successfully applied to masses of material smaller than single bacterial cells, it may seem odd that they are so little used for the general detection of microorganisms. The word sensitivity is often misused, however, and is in any case open to many interpretations. In this chapter, it will be emphasized that the ultimate performance of any analytical tool is of relatively minor importance. What matters is how well it can separate the property of interest from perturbations or fluctuations in the interfering material inevitably occurring with it.

The *limit of detection* defines the true usefulness of an analytical procedure. In many chemical analytical procedures, the limit of detection can be defined precisely with the aid of mathematical statistics.[1] At the same time, however, a great deal of mathematics has been derived in the electronics and communications sciences regarding the detection of weak electrical signals. In many places, the two mathematics are parallel, although the terminologies are different. As a result of the general infiltration of electronics into everyday life there is an increasing tendency to refer to any quantity we wish to detect as the *signal* and the interfering background that hinders its detection as *noise*. The concepts of communication theory have proved so valuable that they are often used in situations where they are less well defined but nevertheless extremely useful in helping us understand what is happening. These concepts are valuable when we attempt to describe the detectability of microorganisms in food.

In electronics, signal strength is generally measured by its rms* voltage, but in other fields, other units are used, e.g. decibels, ergs/sec., hydrolysis rates, lumens/sq. meter. Signals are always accompanied by spurious material (noise), the level of which is described by the same units as the signal. The detectability of a signal has little to do with its actual size; it is almost completely governed by the *signal to noise ratio,* i.e. the relative levels of the signal and the noise in which it is buried. As the signal to noise ratio decreases, the reliability with which we can detect a given signal decreases regardless of the amplifying capability (gain) of our detector because noise is amplified at the same time as the signal.

---

* The root mean square value of a time-variant vector.

Our abilities to detect a distant radio station, manganese in steel, radioactive scintillations in a counting vial, enteric bacteria in salami, a night cyclist's rear light on a busy street, and so on are all governed by the same very fundamental relationships. Unfortunately, there are many instances where the nature of the quantities involved are very poorly defined and only the qualitative workings of the relationships can be seen. Thus, it may be obvious that a motionless, unwinking frog shares many of the visual features of mud, deadwood, and weeds although we would be hard put to quantify them. We can see (qualitatively) why the probability of our detecting a frog at the edge of a pond is low because our eyes are confused by a profusion of similar features, i.e. the signal to noise ratio is low. It would be difficult for us to predict, however, how long we must gaze at the pond before we do see a frog. Incidentally, we can be sure that the frog's detectability will increase enormously as soon as it jumps because the probability of the other pondy features executing ballistic trajectories is very low, i.e. the signal to noise ratio instantly becomes very large.

Regardless of the apparent chemical, biochemical, or immunological bases of methods used to detect the presence of microorganisms (or anything else), we must eventually make use of some physical property such as the evolution of heat, absorption or emission of electromagnetic radiations, ionization current, electrical impedance, and so on. Even if we do not use "instruments" as such, we still rely on physical changes; our eyes, for example, only provide us with information through the changes in light absorption or emission they perceive.

Very responsive detectors are available for almost any of the physical properties one can think of; the problem is that although they can be applied to detection of microorganisms, none of these properties are entirely specific to microorganisms. It is impossible for microorganisms to occur naturally in the absence of materials having (to a greater or lesser extent) the same physical properties. In any specimen, these materials (or rather, the fluctuations they generate in any property we try to measure) represent the noise that determines the method's limit of detection. Quite obviously, the effectiveness with which the analytical methodology separates or isolates the microbial property from the interfering background profoundly affects the signal to noise ratio or the limit of detection. For this reason, it is meaningless to talk of limits of detection on a broad scale, as for example, in *spectrofluorimetry* or *gas chromatography*. Spectrofluorimetry, as such, cannot be described as a technique capable of detecting minute quantities of material. The *limit of detection* applies only to a sharply defined, complete analytical procedure; a small change — for example, the addition of a rinse or filtration step — may change its value considerably.

To better appreciate the effect of noise on detectability, a little mathematics is desirable. In what follows, I have used an argument of a type mainly applicable to chemical or similar analyses since it seems to me that these are more easily visualized. However, electronic terminology is brought in at times to help impress the universality of the relationships.

The aim of a traditional microbiological analysis is to find the concentration c (or number per unit volume or mass) of microorganisms in a food. To the developer of alternative instrumented methods it will be apparent that the analysis must proceed by way of at least one measurable physical quantity x, which may be, for example, a weight, time, extinction, intensity of fluorescence, electrode potential, radioactive count, or an impedance. For any measurement, although it may not be formally defined, there exists an *analytical calibration function:*

$$c = f(x) \qquad (3{:}1)$$

To the microbiologist, who is accustomed to dealing with colonies on plates or altered bottles of broth, it may not be very apparent; nevertheless, the function exists. For example, the measurement property determined by the ability of his eyes to perceive circular outgrowths (colonies) amid particles of food debris on a Petri dish is described by this equation.

In microbiology, c can approach or even equal zero but can never be negative (there is no such thing as a negative microorganism), while the measurable quantity x may take positive, zero, or negative values, depending on the analytical procedure. Because of this, if care is not taken in applying the analytical calibration function, apparently impossible values of c can be obtained at or around the limit of detection.

The limit of detection is a statistical concept, resulting from the existence of an *analytical perturbation level*. At low concentrations, we are never sure whether and by what degree an observed value of x is really due to the material of interest or to uncontrolled chance variations.

Suppose, for instance, we apply the analytical procedure to blank samples (samples believed to contain no microorganisms) and obtain values of x_{bl}. These will not be identical but will exhibit fluctuations owing to, for example, microbial contamination, impurities or other variations in the reagents, adsorption on the walls of vessels, errors of weighing, diluting, etc., secondary reactions, temperature or pH fluctuations, scratches, debris or other optical defects in cells or plates, and so on. The size of the fluctuations is not usually predictable theoretically because we cannot investigate their causes in detail. This fluctuation leads us to the necessity of calculating a mean blank $\bar{x}_{bl}$ and its standard deviation σ_{bl} (the analytical scatter).* This analytical scatter and the electronics term *noise* are equivalent; both are manifestations of the ubiquitous random backgrounds that place limits on the lower level of detectability of any measured property.

We are now faced with the problem of deciding how a measured quantity can be recognized as real or whether it should be rejected as being possibly just a high value of the blank, occurring by chance. Without additional information this cannot be done with certainty. The best we can do is to agree to accept the signal if it is greater than (or equal to) a stated multiple k of the standard deviation, the value of which is determined by the statistical degree of confidence we require.

In electronics a signal is generally reckoned to be detectable if its rms voltage is twice that of the rms noise voltage. It seems reasonable, therefore, for the purposes of this argument to take k = 2. If the fluctuations obey a Gaussian normal distribution

---

* Note that $\bar{x}_{bl}$ and σ_{bl} are only quoted after allowing for all changes that are not due to chance and that can be calculated, e.g. changes due to a different batch of growth medium, a new photomultiplier tube, and so on.

and only values greater than $\bar{x}_{bl}$ are considered, this leads to a confidence level of 97.7 percent in the geniuneness of the analytical signal.[1]

Suppose now we have an analytical signal $x = x_A$. If the sample analysis and blank analysis are made independently, the condition for accepting the signal as genuine is

$$x_A - \bar{x}_{bl} \geqslant 2\ \sigma_{bl} \qquad (3:2)$$

The smallest recognizable measure is given by

$$\underline{x} - \bar{x}_{bl} = 2\ \sigma_{bl} \qquad (3:3)$$

and any smaller measures are not distinguishable from the analytical scatter.

Different types of analysis may provide different forms of Equation (3:2). For example, when sample and blank analyses are paired, the immediate value of $x_{bl} = (x_{A,bl})$ being used as a correction to the particular x_A belonging to it, e.g. as in a double beam spectrophotometer, the criterion for genuineness is[1]

$$x_A - \bar{x}_{bl} \geqslant 2\sqrt{2}.\ \sigma_{bl} \qquad (3:4)$$

In general, writing $\sigma^{\square}$ for the appropriate standard deviation of blank measures (where $\sigma^{\square}$ may be σ_{bl}, $\sqrt{2}.\ \sigma_{bl}$, etc.) the value of $x = \underline{x}$ at the limit of detection is given by

$$\underline{x} = \bar{x}_{bl} + 2\ \sigma^{\square} \qquad (3:5)$$

and the concentration $c = \underline{c}$ at the limit of detection is given by

$$\underline{c} = f(\underline{x}) \qquad (3:6)$$

Sensitivity

At this point, we can properly define the term *sensitivity* because the calibration function may be developed as a series:

$$c = f(\bar{x}_{bl}) + \frac{df}{dx}.2\sigma + \ldots \qquad (3:7)$$

The slope $(\frac{df}{dx})$ of the analytical calibration curve is the true sensitivity of the analytical procedure. It will be seen from Equation (3:7) that the limit of detection $\underline{c}$ of a procedure is quite different from its sensitivity.

The term must, in fact, be used or viewed very carefully outside this analytical context, especially when instrumentation of a technique is being considered. Some authors define sensitivity — being a measure of instrument effectiveness — as the ratio of output to input; others more precisely define it as the derivative of this. In commercial practice it may be defined either as the ratio of output to input under static conditions or as a threshold, i.e. the largest repeatable value of the minimum input change able to produce a definite change of output. This last figure is actually a limit of detection.

LIMITS OF DETECTION OF MICROORGANISMS

We can approach an understanding of the limitations in our ability to detect microorganisms by considering a general example where the microorganisms occupy a volume fraction v_M of the specimen and where the detection system is based on some physical property x. If x_M is the specific value of x for unit volume of microorganisms, the microbial contribution to the total (measured) value of x will be

$$[x_M] = v_M x_M \qquad (3:8)$$

However, the complete specimen will also contain other materials 1,2,i,n, etc., for which

$$v_M + \sum_{i=1}^{i=n} v_i = 1 \qquad (3:9)$$

each of which will possess the property to some extent and will also contribute to the total value of x according to

$$[x_i] = v_i x_i \qquad (3:10)$$

The measured value of x at any instant is thus:

$$x = v_M x_M + \sum_{i=1}^{i=n} v_i x_i \qquad (3:11)$$

The value $\sum_{i=1}^{i=n} v_i x_i$ represents a background which, if it were steady, could be allowed for and would not interfere with detection of the small microbial signal, regardless of its magnitude. (That this is so can be easily seen by imagining a situation in which a large DC voltage is superimposed on the input to a pen recorder. Because the added voltage is constant, we can alter the effective zero point of the recorder until we see only the fluctuations of the signal). Unfortunately, $\sum_{i=1}^{i=n} v_i x_i$ is equivalent to x_{bl} and, as we have seen, this is not constant. Every component i of the specimen experiences perturbations in its environment that cause its contribution to the total value of x to fluctuate. The fluctuations (noise) may be of a thermodynamic nature, i.e. resulting from the thermal motion of atoms and molecules in the specimen, or they may result from sample to sample variations, e.g. in fat concentration, pH, moisture, or color. In general, the magnitudes of the latter will be by far the greatest.

Each contribution to the observed value of x thus exhibits a mean value $[\bar{x}_i] = (\overline{v_i x_i})$ with a standard deviation $\sigma_{[x_i]} = \sigma_{v_i x_i}$. The microbial contribution will, of course, also show a fluctuation, but this can be ignored here.

The measure on which determination of the microorganisms is based is

$$[x_M] = x - \bar{x} \tag{3:12}$$

$$= x - \sum_{i=1}^{i=n} v_i x_i \tag{3:13}$$

and, by analogy with Equation (3:2), the condition for accepting it as genuine, i.e. as actually being caused by microorganisms, is

$$x - \bar{x} \geqslant 2 \sum_{i=1}^{i=n} \sigma_{v_i x_i} \tag{3:14}$$

or

$$[x_M] \geqslant 2 \sum_{i=1}^{i=n} \sigma_{v_i x_i} \tag{3:15}$$

The noisy background thus places a lower limit of detectability on microorganisms:

$$[\bar{x}_M] = (v_M \bar{x}_M) = 2 \sum_{i=1}^{i=n} \sigma_{v_i x_i} \qquad (3:16)$$

Its value will depend on the specificity of the property x on which measurement can be based and the extent to which fluctuations of this property resulting from perturbations in all other components of the specimen that possess it can be reduced.

Depending on convenience, we could write either:

$$\sigma_{v_i x_i} = x_i \sigma_{v_i} \text{ or } v_i \sigma_{x_i} \qquad (3:17), (3:18)$$

i.e. the perturbations could be considered to affect either the apparent volume fraction v_i of a component or the specific value of its property x_i. Thus we may write that, for detectability:

$$x_M \geqslant 2 \sum_{i=1}^{i=n} \frac{v_i}{v_M} . \sigma_{x_i} \text{ or } v_M \geqslant 2 \sum_{i=1}^{i=n} \frac{x_i}{x_M} . \sigma_{v_i}$$

The implications for the detection of microorganisms at concentrations corresponding to conventional regulatory levels can now be seen. The volume of a bacterial cell, for example, is typically about 10^{-12} ml. Statutory levels for pathogenic bacteria tend to range from 100 per g to (less than) one per 100 g. Assuming that bacteria and foods have similar specific gravities, the regulatory volume fractions of these bacteria are in the range of 10^{-10} to 10^{-14}. Substituting $v_M = 10^{-14}$ in Equation (3:19), we see that in the worst case, even if we consider that the whole body of a bacterium contributes to the total x:

$$x_M \geqslant 2.10^{14} \sum_{i=1}^{i=n} v_i \sigma_{x_i} \qquad (3:21)$$

Generally, however, to detect a particular microbial species we must detect some property that is characteristic of those organisms, e.g. an immunologically specific site. Such sites may contribute less than 1/1,000th of the volume of the microorganisms. When we are looking for a particular microbial species, therefore, it is not unreasonable to write

$$x_M \geqslant 2.10^{17} \sum_{i=1}^{l=n} v_i \sigma_{x_i} \qquad (3:22)$$

as the worst case, e.g. *Salmonella* in milk powder.

To detect these microorganisms, then, either the measurement property x must be extremely specific to the cells, *or* we must be able to eliminate it or hold it constant to an unusual degree $(\sum v_i \sigma_{x_i} \leqslant 1\bar{0}^{17})$ in all other materials possessing that property to any extent in the specimen. The best case ($v_M = 10^{-10}$) while 10,000,000 times more feasible than the worst will still be seen to pose very severe problems of specificity or stability.

LIMITS OF DETECTION IN PRACTICE

Describing detectability in this way at least outlines the requirements and suggests that we face a formidable problem when we consider developing rapid instrumental techniques for food microbiology if they are to provide data relevant to existing regulatory levels. However, while we often instinctively make use of the relationships described by Equations (3:19) or (3:20) when we attempt to improve the practical performance of an analytical procedure, we cannot directly use these equations to predict the limit of detection of any procedure because, as has been said, the fluctuations causing the analytical noise or scatter $\sigma_{v_i x_i}$ cannot be investigated in sufficient detail. The value of $\sigma_{v_i x_i}$ can generally only be determined experimentally. We should, therefore, look at some length at current microbial detection limits in order to obtain a feel for the problem. In the examples following, further ideas relating to the detection of microorganisms are mentioned as they seem appropriate.

Before describing the rather dismal limits as they apply to instrumental techniques, however, and to put them in perspective, it should be mentioned that many years ago, *one* property was discovered that is completely specific to microorganisms. This is the property of *multiplication* (or *multiplication rate*). For this property, $^\sigma v_i x_i = \sum v_i x_i = O$. No matter how small its value, the microbial contribution $[x_M] = v_M x_M$ may always be distinguished from the contribution (zero) made by other materials in a food. Multiplication, of course, is the basis of all the successful microbiological techniques — the plate count, most probable number count, and so on. It is the reason for the superiority of the microbiologist's existing techniques and the cause of the Capability Problem described in Chapter 2.

EXAMPLE 1. SPECTROPHOTOMETRY. The spectrophotometer is an instrument of modest capability. In spectrophotometric procedures, it is frequently possible to arrange for the limit of detectability to be controlled by the instrument itself rather than by noise within the specimen. Let us assume that the microorganisms in a specimen can be persuaded to quickly manufacture their own weight of an intensely absorbing substance, e.g. a dye, with a molecular weight M of 100 and a molar absorptivity ϵ of 10^5. By definition, the number of grams m of the dye detectable in volume V of solvent is given by

$$m = \frac{AVM}{10^3} \tag{3:23}$$

Now, on any spectrophotometer, the most accurate readings are given at an absorption value A of 0.3 (about 50% transmittance). Since, as a rule, about 3 ml of liquid are needed to fill a 1 cm cell, under these favorable conditions the spectrophotometer will be able to detect about 9×10^{-7} g of the dye and therefore of the microorganisms. If we further assume that the microorganisms were only diluted by a 1 : 10 factor in releasing them from the specimen into the dye precursor, we see that this spectrophotometric procedure has a limit of detection of about 9×10^{-6} g of microorganisms per g of food, or about 9×10^6 organisms per g.

The value taken for molar absorptivity ϵ was the theoretical maximum value; in practice, one would be lucky to find microbial metabolic products having ϵ as high as 10^4. The limit of detection in this direct functional spectrophotometric procedure, therefore, cannot be expected to be much better than from 10^7 to 10^8 organisms per g of food. It should be noted, however, that by careful attention to the extraction side of the procedure, permitting measurements to be taken in a less accurate region of the spectrophotometer's range, and by extending the duration of contact of organisms with the dye precursor so as to take further advantage of catalytic and/or multiplicative activities, this limit may be somewhat improved. The analytical procedure has then been modified in the manner described under The General Nature of Currently Practical Methods.

EXAMPLE 2. FLUORIMETRY. Closely related to spectrophotometry but generally regarded as being much more sensitive, fluorimetry depends on measurement of the intensity of longer wavelength radiation emitted following the absorption of an exciting beam by fluorescent molecules in the specimen. Since the intensity of the fluorescent radiation I_f is proportional to both the molar absorptivity ϵ *and* the intensity of the exciting beam, it follows that detectable levels of fluorescence may be obtained from smaller and smaller quantities of material simply by increasing the intensity of the exciting radiation. Indeed, quinine sulphate may be detected in water at a level as low as 10^{-10} g by this technique — a considerable advance over the spectrophotometric method. The limit of detection in functional fluorimetry now depends, however, not on the capability of the instrument but on the *blank* fluorescence, i.e. the analytical noise $\sigma_{v_i x_i}$. For most practical purposes, the limit of detection is only 10 to 100 times lower than in functional spectrophotometry.

A number of microbiological procedures have been developed, based on the very specific, i.e. $x_M \gg x_i$, adsorption of fluorescent antibodies to microorganisms. Problems of reagent purity, e.g. the preparation of pure antibodies, a little nonspecific adsorption ($x_i \neq 0$), and persuading microorganisms to manufacture suitable absorption sites on their surfaces, at present provide the unescapable analytical noise that limits detectability. For example, the lower limit of detectability of *Salmonella* cells from foods is about 8×10^5 cells/ml *after* much interference has been removed by extracting the organisms and culturing them in a broth for several hours. There is considerable potential for further development of all immunological methods, and it is probable that they offer the best chances of providing instrumentation related to conventional microbiological concepts.

EXAMPLE 3A. RADIOTRACER METHODS. Radiochemical techniques spring to mind immediately when extreme feats of detection are discussed. Of the multitude of radiation detectors, the scintillation counter, which detects mainly α and β particles produced by radioactive decay, is by far the most widely used and generally thought of as the most sensitive. Consider a typical scintillation counter for which the standard deviation of the background count level (reagent and instrumental blank) may be around 25 cpm,* operating at an efficiency of 15 percent with tritium (^{3}H) labeled material. The presence of tritium in a scintillation vial will be easily recognized if it produces another 50 cpm over the background. A typical commercial tritiated compound, ^{3}H-glucose, has an activity of 6,000 mCi/mM; to produce the required additional 50 cpm requires only 4.5×10^{-12} g of this compound, an amount corresponding to the growth requirements of only one or two bacterial cells. Thus, it would seem that a method based on the incorporation of tritium into a tissue from ^{3}H-labeled glucose might be an excellent way of detecting microbial metabolism at a very low level in food. This is not the case, however, for a rapid natural exchange of labile ^{3}H$^+$ ions with protons in the sample occurs. The rate of exchange is affected by many variables (temperature, pH, water activity, etc.) and is of such a magnitude and variability that any small changes caused by microbial metabolism $[x_M]$ are indistinguishable against it. Its large standard deviation sets a limit of detection many orders of magnitude higher than the ultimate limit of detection of the scintillation counter method for tritium. The rate of incorporation of ^{3}H through microbial metabolism is the signal. The standard deviations of the reagent, instrument, and the much larger (chemical) exchange rate blanks are noise. The noise level in this example may be so high as to prevent detection of microbial metabolism altogether.

EXAMPLE 3B. RADIOTRACER METHODS. Consider a very similar example, using the same scintillation counter, with the same

---

* While the instrumental background may be around 50 cpm in a typical scintillation counter, and its standard deviation (being purely a statistical effect) is calculable, the reagent blank is a less calculable variable. The value $\sigma_{bl} = 25$ cpm taken seems reasonable.

background count level, operating at an efficiency of 85 percent for ^{14}C-labeled compounds. This time, however, we collect the oxidation product $^{14}CO_2$ from the much stabler ^{14}C-glucose. The situation is now much better. A typical ^{14}C-labeled commercial glucose has an activity of 250 mCi/mM; the metabolism of approximately 2×10^{-11} g of this is required to produce an additional 50 cpm over the background in the scintillation counter. While this is a greater mass than in Example 3a., the interference (noise) from chemically exchanged ^{14}C is very small. Noise is not totally absent, however, since radiolytic decomposition of ^{14}C-glucose itself provides a small background variability that sets the attainable limit of detection.

The limit of detection of the ^{14}C-glucose-$^{14}CO_2$ method is sufficiently low, in fact, for it to form the basis of a practical rapid (if not automated) technique for estimating fecal coliform organisms in water.[2] The practical technique requires that cells, after concentration by membrane filtration, receive a two-hour resuscitation incubation, followed by a two-hour incubation with labeled glucose, lactose, or mannose. A detection limit of twenty cells is quoted; this is a little misleading, however, and it must be emphasized that the practical technique is not a direct instrumental method for detecting microorganisms. It is, in fact, one of the compromise approaches discussed at the end of this chapter — approaches where a technique of good limit of detection and a little incubation are combined to minimize the problems from both. During the four-hour total incubation period, cell numbers and the corresponding rates of production of $^{14}CO_2$ can increase exponentially to as much as 250 times their original values. The last few minutes, when cell numbers are at their greatest, therefore, contribute the bulk of the observed metabolic activity. The technique may be described as estimating the existence of twenty fecal coliform cells in a specimen, but detection of their metabolism depends more on the collective activities of some 1,000 to 5,000 cells by the time the analysis is complete.

EXAMPLE 4. BIOLUMINESCENCE. A few years ago, a very powerful method for detecting adenosinetriphosphate (ATP) was developed.[3] The method is based on the ability of ATP to stimulate emission of light from the luciferin-luciferase enzyme-substrate system. It is very selective and capable of detecting femtogram (10^{-15} g) quantities of ATP. It happens that ATP is present in all living tissues in fairly uniform concentrations, it decays rapidly in dead tissues, it can be extracted relatively easily for analysis, and its limit of detectability corresponds to the quantity occurring in a few tens of bacterial cells. Great claims were made at the time by some commercial manufacturers of instruments (who produced delightful photometers with which to measure the ATP-driven luminescence) that the ATP method could be used to determine total numbers of microorganisms in water, urine, milk, and foods.

In the first two cases, these claims were easily substantiated (see next paragraph). However, in the last two, and particularly in the case of foods, the true situation was rather dismal, for foods are (or once were) living tissues. As such, ATP is intrinsic to them, and although it is destroyed when cells die, it disappears neither entirely, instantly, nor uniformly. Remembering that one bacterial cell weighs less than 10^{-12} g, it can be seen that without some means of differentiating the two, the intrinsic ATP of foods must decay by a very large factor before its variability (noise) becomes insignificant against an interesting microbial level of ATP (signal). In fact, using the available analytical procedure, the ATP method was quickly shown to have little ability to detect microorganisms in foods unless they existed at levels of 10^8 or more per g.[4] By the time such levels are reached, most foods already exhibit signs of spoilage.

This example provides a good illustration of the effect of methodology on signal to noise ratios, the trade that must be sought between any benefits a technique may (in principle) possess, and the labor requirements or other complexities needed to make it work satisfactorily. At the time interest was at its peak, the simple extraction procedure (which involved little more than damaging cell membranes using an organic solvent, then filtering out the ATP) was quite incapable of separating the microbial and intrinsic ATP contributions in a food. Later developments, for example, the initial use of ultrasound to isolate bacterial cells from the bulk of many foods, followed by short incubations with apyrases (enzymes capable of destroying free ATP), reduced the contribution from intrinsic ATP and, therefore, lowered the limit of detection of the technique by a factor of 100 or more. It should be noted that the limits of detection for water or urine were already satisfactory because of the ease with which bacteria in them may be filtered out and washed free from intrinsic ATP.

At this level, the technique can still be regarded as a direct estimation of microorganisms. However, a further step (making for a much more practical technique) in which incubation is extended to improve the disappearance of intrinsic ATP while encouraging the microorganisms to multiply, will be seen to have removed the directness of the method, as occurred in Example 3b. In attending to all this extraction and incubation, the microbiologist is apt to find that he spends far more time per sample than he would carrying out, say, a standard plate count. A point may be reached where any advantages due to the rapidity of the method are outweighed by the extra labor involved in carrying it out. This is a common situation in food microbiology. Signal to noise ratios can often be improved to the extent that available instruments may be used to provide alternative data to plate counts, but by then the extra labor required to execute the method often negates its original advantages. The processing may be mechanizable, of course, but its development cost (particularly when it is seen to provide only one of the less valuable items of information required — the total number of organisms) is likely to prevent it from becoming a commercial reality.

EXAMPLE 5. LIGHT MICROSCOPY. Long before scintillation counters or luminescence photometers were developed, microbiologists were already using their eyes as the detectors in a remarkably powerful instrument — the light microscope. When he is sure that the objects in his field of view are microorganisms (as, for example, in checking a culture for the presence of contaminating growths), the microbiologist often has little difficulty detecting and gathering superficial information (size, shape, mobility, gregariousness, Gram-positiveness, etc.) on bacteria themselves. As a detector, therefore, the light microscope/ microbiologist combination has an apparent limit of detection measured in terms of single bacteria. It is natural to ask why the microscope is not, in consequence, greatly used for the routine examination of foods for microorganisms.

The problem is noise again, of course, but the proper description of the limit of detection is also important. The volume in view in a single microscope field at a magnification of 1,000 times is barely 10^{-7} cm^3; to provide a reasonable chance of seeing a microbial cell within a single field thus requires cell concentrations of the order of 10^7 per g or ml. When we consider the light microscope in the context of food microbiology, we should divide this figure by the number of fields (say, 100 to 1,000) the microscopist can reasonably be expected to examine for each food specimen. The limit of detection must thus be considered to be about 10^4 to 10^5 cells per g of food, not a single cell as it at first appeared. This would still be a useful limit if it could be attained. Unfortunately, it rarely can be.

Suppose we placed one stained bacterial cell on a perfectly clean (zero optical noise) microscope slide, somewhere within a 25 × 25 mm square viewing area. It would take the microbiologist but a moment to find it. At 100 × magnification (just high enough to see a dark dot against the light field), his field of view will be about 0.9 mm diameter, or 1/1,000th of the designated area, and by scanning diligently for a minute he will see the cell. Suppose we now provide a little more realism by adding a few specks of fine dust (optical noise) to the slide. The microbiologist must now change his magnification to 1,000 × so as to be able to see the shapes of particles and particularly the one of interest (the signal). At this magnification, his field is barely 1/1,000,000th of the designated area, and scanning the slide will take many minutes. Moreover, since his depth of field will now be barely 0.001 mm, surface irregularities (more optical noise) in the slide will constantly affect his focus. Detecting the microbial signal in the presence of this noise has become much more difficult. If we make things even more realistic by adding a thin layer of ground beef to the slide, the task of finding a bacterial cell becomes almost impossible as thousands of fat globules, mitochondria, and other organelles compete for the microbiologist's attention through the similarities of their sizes, shapes, colors, etc. to those of the bacterium.

In the first (ideal) case, detecting the cell was easy; the microbiologist could be certain that the object he fixed upon was the bacterium, for no other local fluctuations of optical density, i.e. particles, existed. The signal to noise ratio would have been infinite — a pleasant but impossible situation. In the second case, the existence of a little noise led to a finite signal to noise ratio. The microbiologist could never be completely certain that the object he fixed upon was the cell; always there would remain the possibility that he had missed the real cell and chosen a dust particle of the right size and shape.

In the third (very noisy) case, his chances of detecting the cell were extremely low. We could rephrase this by saying that the accuracy of his analysis for bacteria is extremely poor on account of the poor signal to noise ratio. If each field of view contained ten particles capable of being mistaken for the bacterial cell, then his probability of making a correct choice was 10^{-7}. To improve his accuracy to the extent that he would have an even chance of making a correct choice, the slide would need to contain 10^7 bacterial cells similar to the one in question, i.e. the bacterial signal would have to be repeated 10^7 times. If we assume that the slide was smeared with 0.1 g of meat, we see that the contamination level would need to be 10^8 organisms/g to provide him with such an accuracy. This is, in fact, a reasonable, practical example of the limit of detection of the light microscope for bacteria in stained foods.

Miscellaneous Examples and Exceptions

From the examples already described, it should be apparent that analytical noise is the ever present problem limiting the direct detection of microorganisms by instrumental methods, at least at levels corresponding to existing statutory limits. The number of further examples that could be given is enormous, and just a few more should be mentioned very briefly.

The sensitivity of a radiometric deoxyribonuclease assay for *Staphylococcus aureus*, using a radioactive DNA substrate, is limited by the background caused by radiolysis or natural hydrolysis of the substrate to a detection limit of about 10^5 cells/g of this organism. Permutations of methods based on normal or pyrolysis gas chromatography, with flame ionization, electron capture, or even mass spectrometer detectors are limited to the identification of preisolated cultures from carefully defined broths, owing to interference by the complex profiles generated by the foods in which the organisms occur. And even microscopy, using highly selective stains such as the fluorescent *Salmonella* antibodies, suffers an interfering background caused by impurities in the reagents or nonspecific staining.

There are exceptions, of course. To my knowledge, these are all nonspecific methods used for the estimation of total microbial activity or number. The total number of microorganisms in food is not generally subject to statutory limitation, although upper levels are sometimes recommended, and its value may often be noted in assessing the hygienic quality of a food. Total microbial numbers are generally much greater than numbers of specific pathogenic organisms so that direct estimation is a little more feasible. This is particularly so in foods where relatively extensive microbial growth at some stage is not uncommon. Some of the dye-reduction tests in meat[5] or milk,[6] for example, are sufficiently rapid that little multiplication occurs during the period of the test. These are relatively easily instrumented and can, therefore, be considered as being direct instrumental methods of analysis.

Two other routinely used methods that make use of the accumulated microbial activity at the instant of sampling should also be mentioned. Measurement of the oxidation-reduction potential Eh of ground beef, using a commercial Eh electrode system,[7] provides an almost instantaneous datum on meat quality. It is not uncommon, of course, for total microbial numbers in ground beef to be in the region of 10^6 to 10^7 organisms/g. Measurement of pyruvate (a common microbial metabolic product) in milk, using an automatic chemical analyzer,[8] provides direct information about the accumulated metabolic activity in the specimen. It is interesting that pyruvate, being stable to heat, will survive to provide information on the history of milk long after the organisms themselves have perished in the pasteurization process.

There are a few more examples like these, but their number is small compared with the great range of our interests in food microbiology. The most important of these interests — determination of specific pathogenic organisms, as these are described by, and at the levels defined by statutory standards, guidelines, or recommendations — are at present inaccessible to direct, instrumented analysis. Moreover, we should not hold out too great hopes for dramatic advances in the foreseeable future. The problems lie not so much with instruments themselves as with techniques for separating substances of microbial origin from the overwhelming quantities of food in which they occur. Improvements do occur from time to time. We should keep close eyes on, for example, the radioimmunoassay based analyses and, in particular, the newer enzyme-linked immunosorbent assays for direct detection of microbial toxins. However, in many instances, determination of microorganisms or their products at levels pertinent to current statutory standards requires capability jumps of many orders of magnitude. We have a long way to go.

THE GENERAL NATURE OF CURRENTLY PRACTICAL METHODS

Faced with a need for rapid instrumented methods of determining microbial numbers, many inventors have resorted to compromise approaches, employing abbreviated incubation (microbial multiplication) stages to improve signal to noise ratios. Some practical examples of these were mentioned in Examples 2, 3b. and 4. The rationale behind these compromise procedures is described below. It will be seen that there are some problems; however, in general, they seem to hold the most promise for instrumenting food microbiology over the next few years.

Let us assume that the microorganisms of interest produce some metabolic product for which we have suitable detectors but that, unfortunately, we cannot measure directly in foods owing to the magnitude of the analytical noise level. We may, for example, be able to detect minute changes in electrical conductivity or impedance, but the magnitude of the changes in this property caused by reasonable levels of microorganisms in foods may be lost in the natural variations occurring either through specimen to specimen variation or within a specimen as a result of intrinsic chemical or biochemical processes.

Suppose now we extract the microorganisms from a specimen and transfer them to a carefully defined growth medium in which they can begin to produce impedance changes. We still find an analytical noise level, somewhat reduced compared with the food itself but nonetheless existing as a result of perturbations caused by the introduction of the extract, dissolved oxygen or other impurities or variations in the medium, temperature fluctuations, and so on. If the concentration of microorganisms is sufficiently high, we will quickly, i.e. before significant multiplication occurs, observe an impedance change that may be confidently ascribed to their metabolic activity.

Let us assume that the growth medium has an average impedance value $\bar{z}_{bl}$ with a standard deviation $\sigma_{z_{bl}}$. (Strictly, we should not expect these to remain constant with time; however, without knowledge of the forms of the expressions involved, there is little point in adding this extra complication to the later argument, and we shall assume that $\bar{z}_{bl}$ and $\sigma_{z_{bl}}$ remain constant.) We shall also assume that the rate of contribution of each and every cell $(\frac{dz_M}{dt})$ to the total impedance change is constant (k_{z_M}) throughout its life.

After time t a concentration c_M of cells will produce a measured impedance change:

$$z - \bar{z}_{bl} = c_M \int_0^t \frac{dz_M}{dt} = c_M t k_{z_M} \qquad (3:24)$$

The condition for accepting this as due to microorganisms [Equation (3:2)] is

$$c_M t k_{z_M} \geq 2\, \sigma_{z_{bl}}$$

We see from this that provided our assumptions about the constancy of $\bar{z}_{bl}$ etc. hold, the minimum detectable concentration of cells depends inversely on the duration of the analysis up to a time when c_M itself begins to change significantly as a result of the tendency of microorganisms to multiply. Up to this point, the method may be described as a direct measurement of microbial metabolic activity in the specimen. Its value as an estimator of microbial *numbers* depends on the distribution of the possible values of k_{z_M} among the organisms we are likely to encounter and the frequency with which individual species are likely to occur. The developer of any such method will produce graphs or tables showing the statistical relation between his data and microbial numbers (as determined by plate counts, for example) in typical situations.

If the analysis is allowed to continue longer, C_M will increase, i.e.

$$(C_M)_t = (c_M)_o f_M(t) \tag{3:25}$$

and the impedance change will accumulate at an increasing rate. Often, $f_M(t)$ is not simple. However, if conditions can be arranged so that all of the cells begin to multiply immediately in the growth medium, and provided that the analysis ends before concentrations of any limiting nutrients or metabolites become critical, the impedance change will increase according to

$$z - \overline{z}_{bl} = k_{z_M}(c_M)_t = k_{z_M}(c_M)_o.2^{t/T} \tag{3:26}$$

(T being the time required for the microbial population to double.)

The minimum incubation time required to detect the microorganisms will be

$$\underline{t} = T \log_2 \frac{2\sigma_{z_{bl}}}{k_{z_M}.(c_M)_o} = 1.44 \ T \ln \frac{2\sigma_{z_{bl}}}{k_{z_M}.(c_M)_o} \tag{3:27}$$

In a situation where multiplication occurs, therefore, this logarithmic relation allows for disproportionately small increases in the time required for detection as the concentration of microorganisms it is required to handle in a specimen decreases. The electrical impedance method itself, for example, often allows bacteria to be detected more or less immediately at concentrations of 10^6 to 10^7 cells/ml. When used in this indirect manner, i.e. when incubation is prolonged to allow multiplication to occur, the procedure often allows detection of bacteria from foods at levels of 10^4 to 10^6 cells/g in five hours.[9]

Nearly all the currently available and reasonably practical instrumental methods of detecting microorganisms in foods are based on this compromise approach of combined procedures. Alone, both contributing procedures have severe drawbacks.

Those conventionally based on multiplication alone (plate counts, etc.) tend to require unduly lengthy incubation periods because the unaided human eye requires the presence of 10^8 to 10^{10} organisms before it can detect them reliably. Those based on direct instrumental determination of physical, chemical, or other properties, e.g. adsorption of fluorescent antibodies, accumulation of ATP, metabolism of radioactive substrates, changes in oxidation-reduction potential, impedance, light scattering, hydrogen ion concentration, while considerably better than human eyes, generally still have limits of detection too high for practical use. Combining the two types of procedures allows microorganisms themselves to provide an adequate signal to noise ratio for instrumental detection within a more acceptable analysis period.

Disadvantages still remain or are introduced by the combination, however. One has already been mentioned; this is the probabilistic relation existing between microbial numbers and any other measured property. It results from the enormous range of metabolic activities and microbial compositions, not just between species and strains but also for individual cells of different histories or under different environmental conditions. Wherever microbial data are required in the form of numbers, therefore, a certain suspicion is always lent to this type of data. The psychological problem experienced by would-be developers in the face of the relative confidences we apperceive in the two types of data, for example:

Plate count
procedure: The sample contains 5×10^5 organisms/g

Instrumental
procedure: There is a 70 percent probability that the sample
                contains 5×10^5 organisms/g

is discussed in greater detail in Chapter 7.

In these combined procedures, this psychological disadvantage is compounded by the introduction of two other variables. The growth media employable in such instrumental methods are usually not those that have been developed for conventional enumerative procedures. Whenever mixed flora occur (which in foods is nearly always), there is thus some doubt about whether the organisms that *do* multiply are the ones that would have multiplied in the conventional method. To this must be added doubts resulting from the distribution of multiplication rates likely to be found across species, strains, and individual cells. The uncertainty that is introduced increases the greater the difference between the limits of detection of the direct and combined procedures.

CERTAINTY FROM THE EXTREMES OF UNCERTAINTY

It would seem from the previous discussion that the most rewarding developments in analytical procedures will be those that improve instrumental limits of detection since these will reduce the difference between direct and combined procedural limits and the uncertainty it engenders. This is the almost instinctive approach of every would-be developer. It may at first seem illogical, therefore, to spend much time investigating the advantages of working in exactly the opposite direction, that is, of seriously considering the potential of combined procedures using currently available detectors at levels far below their limits of performance.

However, as so often happens, going to the extremes of an apparently undesirable lane may lead to previously unnoticed attractions, particularly if the original direction was not unequivocably the most desirable. The production of numerical data — or their probabilistic equivalents — is implicit in all of the methods described. If any doubt happens to be cast on the value of enumerative methods, then the possibility that diametrically opposite procedures and data may be preferable opens. To close this chapter, therefore, we should look briefly at the effects of introducing more and more uncertainty into a combined procedure by detecting microorganisms at ever increasing rather than decreasing levels. When one is dealing with probabilities and correlations, all sorts of things may happen.

Consider just one example. There are numerous approaches we could use in developing instrumental methods for obtaining evidence about the presence of *Staphylococcus aureus* in foods. For example, we could

a. mechanize plate count methods, i.e. methods of directly determining numbers of characteristic *S. aureus* organisms. This property has some statistical relevance to the national level of cases of staphylococcal intoxication, and the measurement is directly pertinent to statutory numerical standards.

b. mechanize methods based on deoxyribonuclease, coagulase, lecithinase, and other activities. These properties when quantized, i.e. into present or absent, show some correlation with one another. They are less relevant to cell numbers and therefore to standards.

c. mechanize methods based on enterotoxigenic activity. This property shows some correlation with all these other properties and is very pertinent to intoxication.

None of these properties correlate perfectly with one another. The reasons for this will be examined in more detail in Chapter 4. Suppose we choose approach (c), the development of instruments based on the determination of staphylococcal enterotoxins. In a typical immunological procedure, we might reasonably achieve a limit of detection of around 0.01 μg of the various staphylococcal enterotoxins per g of food. A typical statutory limit may be around 100 *S. aureus* cells per g, however, and the level of enterotoxins normally capable of being produced by this concentration of cells is considerably lower than our limit of detection. It is obvious that the procedure is not capable of directly providing information on *S. aureus* contamination unless a food grossly violates the standard. The procedure must be combined with an incubation stage in order to improve the signal to noise ratio.

By the time the limit of detection has been modified by incubation to approach the standard, the probabilities involved in deducing cell numbers from enterotoxin levels, the range of rates of multiplication of organisms, and the possibility that other cells have grown up also to inhibit or stimulate the production of toxins are multiplied until there is considerable doubt about the reliability of the data. The uncertainty, however, only relates to whether the original specimen conformed to a standard based on microbial numbers. In the sense of indicating whether or not the organisms are capable of producing at least

0.01 μg per g of enterotoxins during the analysis period, the data are unequivocal.

Now suppose that we change the limit of detection of enterotoxins from 0.01 to 0.1 μg per g of food.* This higher limit will obviously require a longer incubation period. Our uncertainty about whether or not the food conforms to a numerical standard will increase further. However, in the sense of indicating whether or not the organisms are capable of producing at least 0.1 μg per g of enterotoxins during the analysis period, the data will still be unequivocal.

The ingestion of 10 to 100 grams of food containing this higher level of enterotoxins would make an appreciable percentage of humans ill.† In the sense of indicating whether or not the organisms in the food are capable of producing sufficient toxins during the analysis period to make an appreciable percentage of humans ill, the data are, therefore, unequivocal also.

Finally, suppose we modify the analysis procedure so that we measure the enterotoxin concentrations every hour or every five hours, say, until we can detect this level of 0.1 μg per g. We are now in the unusual position of being able to provide a very meaningful item of (legally) rather useless information. We can say, quite unequivocally, that the food can generate a level of enterotoxins capable of causing illness in many people after, say, fifteen hours at our incubation temperature. To most consumers, this is a most interesting and meaningful piece of information about the food. However, we have introduced so many uncertainties by our lengthy incubation period that we cannot say whether or not the food conforms to the numerical standard.

The unlikely path is, thus, not completely barren. Raising detection limits relieves the performance requirements on instruments and techniques, making them more feasible. And raising the limits to a suitable level can lend the data a great significance to humans. To would-be developers of micro-

---

* One can arbitrarily raise the action level of an established procedure to any value. Alternatively, a (generally simpler) less effective procedure may be substituted.

† The quantities of staphylococcal enterotoxins implicated in cases of illness tend to be in the ranges 1 to 4 μg for enterotoxin A, 20 to 25 μg for enterotoxin B, or in the range 10 to 13 μg for enterotoxins A, B, and C, depending on the report.[10]

biological instruments and to those of us who eat food (a not easily dismissable proportion), its attractions may appear worth greater examination. A great deal more can be said about this approach. It will be discussed in far greater detail in later chapters, for the consequences of accepting it as a basis for assessing food quality would be far-reaching.

REFERENCES

1. Kaiser, H.: *The Limit of Detection of a Complete Analytical Procedure.* London, Adam Hilger Ltd., 1968.
2. Reasoner, D. J. and Geldreich, E. E.: Rapid detection of water-borne fecal coliforms by $^{14}CO_2$ release. In *Mechanizing Microbiology,* (Eds.) A. N. Sharpe and D. S. Clark. Springfield, Thomas, 1978, p. 120.
3. Addanki, S., Sotos, J. F. and Rearick, P. D.: Rapid determination of picomole quantities of ATP with a liquid scintillation counter. *Analyt, Biochem, 14:*261, 1966.
4. Sharpe, A. N., Woodrow, M. N. and Jackson, A. K.: Adenosine-triphosphate (ATP) levels in foods contaminated by bacteria. *J Appl Bacteriol, 33:*758, 1970.
5. Holley, R. A., Smith, S. M. and Kempton, A. G.: Rapid measurement of meat quality by resazurin reduction. I. Factors affecting test validity. *Can Inst Food Sci Technol J, 10:*153, 1977.
6. American Public Health Association: *Standard Methods for Analysis of Dairy Products,* 12th ed, New York, A.P.H.A. Inc., 1967.
7. Brown, L. R. and Childers, G. W.: A rapid method of estimating total bacterial counts in ground beef. In *Mechanizing Microbiology,* (Eds.) A. N. Sharpe and D. S. Clark. Springfield, Thomas, 1978, p. 87.
8. Tolle, A. W., Heeschen, H., Wernery, H., Reichmuth, J. and Suhren, G.: Die Pyruvatbestimmung ein neur Weg zur Messung der Bakteriologischen Wertigkert von Milch. *Milchwissenschaft, 27:*343, 1972
9. Cady, P.: Progress in impedance measurements in microbiology. In *Mechanizing Microbiology,* (Eds.) A. N. Sharpe and D. S. Clark. Springfield, Thomas, 1978, p. 199.
10. Gilbert, R. J.: Staphylococcal food poisoning and botulism. *Postgrad Med J, 50:*603, 1974.

Chapter 4

THE SHORTCOMINGS OF MICROBIOLOGICAL COUNTS

. . . the most important fact illustrated by this study is that Standard Plate, coliform, or psychophilic counts on fresh commercially processed milk are poor criteria of potential keeping quality. . . . Those who promulgate standards and laboratory procedures appear to be more interested in the performance of tests according to accepted procedures and compliance with bacterial standards having dubious meaning than they are with applying methods proposed for measuring milk quality at the consumer level. . . .[1]

It is concluded that bacterial counts on raw meats cannot serve as indicators of: (a) health hazards, (b) insanitary conditions, (c) product spoilage, or (d) an aesthetic value of the food. . . . Bacterial numbers per se do not — indeed cannot — reflect the degree of hazard involved in the consumption of the meat, and therefore their application cannot actually increase the protection of the consumer against a health hazard.[2]

. . . indeed, all eight initial bacterial counts combined accounted for only 14 to 31% of the observed variability in days to go bad at any of the storage temperatures. . . .[3]

Negative responses [in a survey of state governments towards microbiological criteria for foods] basically questioned the concept of relating microbial counts to product quality. . . .[4]

E NUMERATIVE MICROBIOLOGY is a quite unique science, both in the manner of the analytical procedures it encompasses and in the way its data are applied to the very objective for which they are obtained. The suggestion was made in Chapter 2 that the significance of microbial counts is illusory, going far beyond the information actually carried in them. This comment is amplified in Chapter 7 where the psychological image of the count is discussed at length. In this chapter, I shall discuss scientific aspects of the count — its value as a scientific datum, as a predictor of food quality, and particularly as an objective for duplication, emulation, or simulation by alternative instrumented methods of microbiological analysis. This last is very important; since both straightforward mechanization or de-

73

velopment of alternative analyses yielding correlating data appear to be most difficult feats, we should closely examine whether the effort to eventually do so is going to be worthwhile.

INDETERMINACY*

We tend to look on numbers as being absolute. To many nonmicrobiologists, at least, this image persists for statements of microbial numbers. The count is a concept every person can grasp, and it is very easy to grasp it too firmly. The following quotation, for example, while actually referring to water microbiology, is equally descriptive of the situation for foods:

> One important aspect of microbial standards . . . is that they provide a common base or language for relating numbers of organisms to the level of quality that is desired. This standardized numerical language is important to the legal people who are responsible for enforcement of water quality compliance regulations. They often treat (or mistreat) these numbers as though they were cast in bronze, probably one of the occupational hazards of their profession.[5]

The relation between plate counts or other enumerative data and the numbers of microorganisms actually in a food is, however, very poorly defined. Microbiologists know, for example, that foods generally contain many more microorganisms than they are able to detect. There are many reasons why enumerative methods fail to provide absolute data. Leaving aside peripheral problems such as the uncertainties about microbial growth or survival while samples are transported to the laboratory and the difficulties of obtaining representative samples from large batches of food, one important reason is seen to be the coherence of microorganisms. Many organisms (bacteria and molds particularly) exist in clusters or chains of viable units, which resist complete dispersal to single cells in the blending processes used in most microbiological analyses. The colonies subsequently formed on agar plates may arise from nuclei of one, two, twenty, or more cells, in a distribution dependent on the food, the microbial flora, and the blending process. There is

---

* I had difficulty finding a word to cover the elusiveness of "hard" data in food microbiology. Webster's New 20th Century Dictionary defines *indeterminacy* as — the property of having inexact limits, indefinite, indistinct, vague, unsettled, undecided, inconclusive. This did not perfectly describe what I have in mind but was preferable to *circumstantiality* (the quality of being modified by circumstances), *contingency* (occurrence depends on chance or uncertain conditions), or *subjectivity* (unauthenticable).

no way of deciding how many cells originated any particular colony.* Thus, in order to be able to approach a consistency in counting, even for a given growth medium, the conditions of dispersal of the microorganisms must be carefully specified. Deviations from this result in lower or higher apparent levels of microbial contamination being quoted.

Even more important, however, is the inability of microbial counting media to support the growth of every organism the microbiologist would like to count. No growth medium (the food included) provides perfect conditions for every organism of interest. The environment of the cells greatly affects their ability to multiply and, therefore, their competitiveness relative to other components of the flora. Being ripped by blender from the true site of competition (the food), cells find themselves dumped into a new environment (the growth medium). The result of the ensuing free-for-all in this artificial territory decides the perceived level of contamination in the food. Implicit in the result is the potential of the food to become unwholesome in storage as a result of microbial growth and competition within it.

Damaged, dormant, or vegetatively virile, a proportion of the cells find the growth medium to their liking while others do not. The count and the great variability it exhibits reflects this. The microbiologist strives to find media yielding the highest possible count for particular groups of organisms regardless of whether or not all of them can actually grow in the food. From the purely bacteriological point of view, this is logical since ". . . results of plate counts and MPN tests are extraordinarily variable between laboratories, and even between workers in the same labora-tory."[6] Greater consistency in counting can presumably be achieved if growth media allowing 100 percent of the organisms of interest to grow are discovered. Unfortunately, as sites for competition, such growth media bear little resemblance to the original foods; their ability to indicate what will actually happen in the food in storage thus suffers.

As a result of searches for "best" growth media, we are treated in microbiological journals to monstrosities such as

---

* Microbiological counts are, of course, usually quoted as numbers of colony-forming units (CFU) rather than cells to accommodate this. Unfortunately, such counts often tend to be referred to and treated as "true" counts whenever comparisons are made with alternative methods of analysis.

The most reliable medium in the overall analysis was mannitol salt agar. However, this medium was not equally reliable at all times during ripening, and use of both mannitol salt agar and Staphylococcus medium no. 110 is recommended. The tellurite- and azide-based selective media were generally unsatisfactory, however, tellurite glycine agar, Vogel Johnson (VJ) agar, and azide blood agar base were totally unreliable. In general, the salt-based selective media were most reliable. This applied also to the egg yolk media that use salt as the selective agent. Salt egg yolk agar and Colbeck's egg yolk medium generally gave higher recoveries of *S. aureus* than did Baird-Parker medium, Crisley et al., tellurite polymyxin egg yolk agar, and Hopton egg yolk azide agar, except in the unripened cheeses. The debilitating effect of cheese ripening on the staphylococcal cells was not eliminated by the egg yolk tellurite and azide media.[7]

I must emphasize that this type of report is by no means uncommon and that my criticism is aimed not at the writer but at the incongruity of a scientific subject that makes such extravagant and expensive comparisons seem necessary. In fact, the report concerns sound and valuable research within the accepted conceptual framework of food microbiology.

The importance of growth media and other analytical factors to the perceived count will be seen when it is considered that very small preparational variations, e.g. resulting from position in the autoclave or exposure to oxygen, for plate count media in different laboratories may easily affect microbial growth rates by ± 10 percent. For example, relative doubling times of 18 and 22 minutes (in the exponential growth period) for two closely related organisms may easily be reversed by small variations in thermal degradation of vitamins, oxidation of cysteine, hydrolysis of proteins, or pH changes occurring simply during sterilization of the media. If the relative rates are maintained throughout colony growth, and if we assume colonies of the two organisms to be just detectable by the human eye when they contain 10^8 cells, it requires but a moment's calculation to find that when colonies of one strain have reached a countable size, the others have only produced about 3.5×10^6 cells and are, therefore, likely to be missed in counting. If the two organisms happen to be fairly uniformly distributed in a food, it is not at all impossible for technicians in different laboratories to arrive at similar counts, although their results may relate to different organisms.

Exactly the same considerations apply to other possible perturbations in growth conditions. The number of variables able to affect relative microbial multiplication rates is so large (they can rarely be completely identified, in fact) that only the most obvious ones can be effectively controlled, and complete standardization in microbiological analyses is virtually impossible.

The resulting fluidity is fundamental to enumerative procedures; it will remain whether or not the procedures are eventually mechanized. At present, of course, microbiologists must also endure human variabilities, such as those resulting between analysts in weighing, pipetting, and particularly in identifying colonies or altered broths. These last are very subjective. Even for a simple food such as milk, bacterial estimates from different analysts counting the same plates have been found to vary by as much as 565 to 948 so that even the *standards of accuracy* specified in Standard Methods for the Examination of Dairy Products[8] have been criticized as being unrealistic.[9]

The capriciousness of microbial counts becomes more and more pronounced for the lower counts on selective media, where increasing demands are placed on the biochemical ingenuities of microorganisms, and on analysts in discriminating between "positive" and "negative" situations. A recent International Commission on Microbiological Specifications for Foods (ICMSF) study of "confirmed coliforms" using most probable number techniques was very revealing.[10] In this study, up to fifteen laboratories examined the same foods, artificially inoculated with coliform organisms at a central laboratory, using centrally packaged media. Variations in laboratory average counts ranging from 7.7:1 for a meat meal subject to 2,290:1 for a nonfat dry milk were recorded.

Of course, provided analytical conditions remain reasonably constant, changes in count for a food can be useful in describing its changing microbiological quality. The quality control staff in a food factory, routinely carrying out the same analysis, for example, may validly use counts as indicators of changing hygiene practices in the factory. However, changes in staff, laboratory facilities, or sources of reagents can upset seemingly established patterns.

Ultimately, the value of making analyses for particular organisms can itself be challenged. The organisms of the world are now known to share a common plasmid pool from which (given time) almost any species or strain may acquire almost any biochemical activity it needs to refit itself to changing environments. If the genetic sequences responsible for, say, the ability of organisms to colonize or invade a host or produce toxic metabolites may pass relatively freely between organisms, the classical characteristics of genus and species become less important.

Enumerative microbiological data are thus very indeterminate and will remain so for the foreseeable future despite continued efforts to develop perfect growth media, i.e. media recovering every organism of interest and no others. Indeterminate data are always questionable. They are apt to lead to conflicting results, as has been seen, and are a poor basis for standards of any kind. The lack of consistency they engender has been a formidable barrier to the introduction of microbiological standards for many foods.

INFORMATION DEFICIENCIES

Before discussing how well the information provided by the more common microbiological analyses compares with the kind of information consumers need for their protection, it will be useful to briefly introduce two concepts that will become increasingly important in Chapters 5 and 6 and that will be described in greater detail there. These are the *parameter of unwholesomeness* and the *threshold of physiological response.*

A food is not and does not become unwholesome because it contains microorganisms per se but because the potential for evoking undesirable physiological or psychological responses is generated within it. The distinction is important, particularly in the cases of organoleptic or toxic responses, although it may seem to be splitting hairs a little in the case of infective agents. Human bodies respond only to physical, chemical, biochemical, immunological or transcriptional* stimuli, and it is only through these that we become aware of the wholesomeness or otherwise of a food. Regardless of the numbers and kinds of microor-

---

* The word is used here so as to separately classify our sensitivity to the virological properties of a food. In fact, while it refers to the ability of virus particles to exploit human cells to their own ends, the eventual parameters by which we become aware of virus invasions are still physical, chemical, biochemical, and immunological.

ganisms in it, a food cannot be considered unwholesome until it can evoke undesirable responses in the consumer. A food containing many cells of *Staphylococcus aureus*, for example, although it may violate a guideline or standard, only becomes unwholesome (in the context of staphylococcal poisoning) as a result of suitable quantities of enterotoxins being generated within it before it is eaten. Thus, the parameters of unwholesomeness are not microorganisms or their numbers but are factors that in other sciences we are accustomed to measuring with a variety of instruments.

Our responses to toxic or infective parameters are very much at the somatic level; our response to organoleptic parameters is partly instinctive and partly conditioned. In all three cases, however, the involvement of microorganisms is important to us only as an associative concept resulting from our education. We do not detect lactic acid bacteria in milk or weiners, but we become aware of them through their effect on the pH in our mouths. In the case of souring, therefore, the total acid and the pH of the food are the relevant parameters of unwholesomeness. Similarly, we do not detect *Staphylococcus aureus* bacteria, but we become aware of them in ham through the eventual effect of their toxins on the vomiting centers of our brains. In the case of staphylococcal intoxication, therefore, the concentrations of the various toxins are the relevant parameters of unwholesomeness. In neither case do we need to be familiar with the concept of microorganisms to experience their effects.

The value reached or *amplitude* of a parameter determines the probability of our becoming aware of it over the normal physiological and psychological fluctuations we experience in life. For any parameter of unwholesomeness there is a *threshold* amplitude, below which we will be unaware of its existence. This is the limit of detection of the parameter by the human body; it is subject to the same signal to noise limitations as any other detection procedure. Since a food cannot be considered unwholesome unless it evokes undesirable responses, it follows that it also cannot be considered unwholesome unless the amplitude of any parameter of unwholesomeness is greater than the consumer's relevant threshold of physiological response.

With this in mind, we can now look at the performance of microbiological analyses in providing information relevant to human physiological responses.

Consider a tasty and wholesome morsel of food. It is wholesome because its physical, chemical, biochemical, and immunological properties have not been so altered that it: (a) is organoleptically unacceptable, (b) produces toxic symptoms, or (c) causes infection.

The food may contain microorganisms, but our senses cannot detect them. Indeed, no useful comments or predictions can be made about its microbiological quality or potential without some kind of microbiological analysis being made.

If the food is stored too long, we can be sure that its quality will deteriorate until we eventually have to describe it as unwholesome. Microorganisms will probably play a large part in this deterioration. At different stages, different parameters of unwholesomeness may become evident; it is even possible that some of the parameters may decrease into unimportance again at times. At any stage, the food may or may not evoke responses. It may repel by its appearance, odor, or taste, or cause consumers to vomit, suffer diarrhea, nausea, cramps, paralysis, blindness or rashes.

> The rest are . . . sharp belchings, fulsome crudities, heat in the bowels, wind and rumbling in the guts, vehement gripings, pain in the belly and stomach . . . much watering of the stomach, and moist spittle, cold sweat, importunus sudor, unseasonable sweat all over the body . . . midriff and bowels are pulled up, the veins about their eyes look red, and swell from vapours and wind . . . their ears sing now and then, vertigo and giddiness come by fits. . . .
>
> Robert Burton, *Anatomy of Melancholy.*

At no stage will our bodies detect microorganisms,* although we may be painfully aware of their activity.

Microbiological analyses may show apparent differences between the stored food and the original. On the other hand, they may not. For example, it is quite possible for weiners to deteriorate in a variety of ways (through the development of surface slime, turbid exudates, sourness, texture changes) without any apparent correlations with counts of different types of organisms being evident or even without there being any apparent

---

* It is arguable whether seeing colonies such as mold outgrowths is an actual example of detecting microorganisms.

change in numbers. Whatever it is that changes, therefore, to make us aware of microbial activity, is not just the number of microorganisms.

Between any arbitrary instant when the food is wholesome and some later hour, at least five variables are important *and must be considered as being equally important* in determining whether or not the food will still be wholesome. For any given parameter of unwholesomeness these are

1. The number (strictly, the concentration) of relevant microorganisms initially present. Generally, the greater their concentration, the stronger we expect their effect to be.
2. The extent to which the relevant organisms can multiply in the interval. Multiplication rates may vary greatly with time as a result of changes in available nutrients, the atmosphere, pH, Eh, degree of proteolysis, etc.
3. The specific cellular metabolic activity of the organisms in producing the unwholesome parameter. This may vary greatly with time throughout the growth phases of individual species. To compound the variability, however, a plurality of relevant species may be present, each having its own characteristic pattern of metabolic activity, and all of the patterns may be perturbed differently by the environmental changes mentioned in (2).
4. The effects of the remaining flora of the food on (2) and (3). While they may not directly contribute to the growth of a parameter, other organisms can exert a profound influence on those that do, as a result of, for example, competition for nutrients, inhibition or stimulation of growth or metabolic processes through chemical or physical modification of the environment, or the activation, neutralization, lysis, or metabolism of their physiologically active products.
5. The consumer's physiological threshold, compared with the amplitude the parameter attains as a result of the concerted action of variables (1) through (4). If during the interval, the combined, integrated variables yield an amplitude below the consumer's threshold, we must say that the food is still wholesome. If it is above, the food is unwholesome, and the consumer can expect to experience an undesirable physiological or psychological response.

Counting microorganisms fixes the conventional approach to assessing food quality as *deterministic*. That is, the behavior of the system is assumed to be describable or predictable from a knowledge of the variables affecting it. However, to be able to determine whether a food will become unwholesome within a given interval requires information on all five of the above variables. It would appear that the ubiquitous microbial count provides only one-fifth of the required information since it produces no information on the remaining variables. In fact, as will be seen later, it may not always provide even this small offering. The outcome of the system — food plus consumer — it should be noted, is *not* a deterministic microbiological conclusion.

Microbiological information is only complete, precise, and relevant to the wholesomeness of food when it satisfactorily predicts the consumer's response. No other information, although it may be easy to obtain, repeatable, apparently objective, or whatever, has such relevance. Let us now look at the information generated by some common methods of assessing microbiological quality and, in particular, how these relate to the information required (or generated, if he is so unfortunate) by the consumer.

As can be seen from Table 4-I, only the consumer sums *all* the variables to provide complete and precise microbiological data. Thus, according to whether or not he/she becomes ill, information is obtained on how the interaction of the number of organisms, their specific metabolic activities, multiplication rates, and the effect of other organisms during the storage interval generate parameters of unwholesomeness relating to his/her threshold of response. However, the consumer may feel that this is a rather poor way to make the analysis.

Unfortunately, only one other microbiological test — using organoleptic assessment panels to rate incubated food samples — produces the same information as the consumer, within its limited range of application. Its realism depends, of course, on simulated consumers. However, its routine use is restricted to low-hazard situations. Examination of cans for swelling during incubation yields information combined from four of the five variables. While it does not yield information on human physiological thresholds, swollen cans do provide unequivocal evi-

TABLE 4-I

RELATIVE QUANTITIES OF INFORMATION PRODUCED BY DIFFERENT METHODS OF ASSESSING FOOD MICROBIOLOGICAL QUALITY

Information	No. of Organisms	Specific Metabolic Activity	Rate of Multiplication	Effect of Other Flora	Human Physiological Threshold
Method of analysis					
The consumer	1	1	1	1	1
Incubation and organoleptic assessment	1	1	1	1	½[a]
Swelling on incubation (cans)	1	1	1	1	0
Enterotoxin determination	1	1	0	0	0
Ammonia	1	1	0	0	½[b]
Extract release volume	1	1	0	0	0
Dye reduction	1	1	½[c]	0	0
Pyruvate (milk)	1	1	0	0	0
Oxidation-reduction potential	1	1	0	0	0
ATP	½[d]	0	0	0	0
Plate count	½[d]	0	0	0	0
MPN count	½[e]	0	0	0	0
Presence after enrichment	0	0	0	0	0

1 = produced; 0 = not produced; ½ = produced under some circumstances. a = not universally applicable; b = has organoleptic relevance if applied to headspace of sample; c = for prolonged incubation and multiple examinations; d = only if organisms disperse easily and grow readily; e = to the extent that organism occurs in sample of given size only.

dence of faulty canning. The range of application of the method is again rather restricted. The remaining types of analysis yield considerably smaller amounts of information, and their values as predictors of unwholesomeness in the samples to which they are applied are correspondingly lower.

In particular, we can see from Table 4-I why the plate count is truly unique. It provides no information on human physiological thresholds or on the relevant metabolic activities of the organisms. Plating conditions are carefully arranged to separate organisms as much as possible, thereby avoiding producing information on interactions between components of the flora. And it does not even reliably provide information on the number of organisms, only the number of dispersible clusters. The most probable number technique is (in principle, at least) capable of providing information on the interaction of the flora because organisms are thrown together in tubes of broth. However, since MPN counts are almost invariably carried out using selective media, any relevance this information may have to the food itself is lost. The MPN count is, therefore, indistinguishable from the plate count in this respect. Both yield very little information. At best, when organisms disperse easily and completely in the blending process, these counting methods provide 20 percent of the required information. Generally, however, they provide less.

The incompleteness and the varying coverage of the required information provided by the various methods of microbiological analysis are the sources of the frustrating inability of all of them to yield satisfactory intercorrelations. When data do not relate to similar combinations of these items of information, any observed correlations can be, at best, a matter of luck. For example, within any analytically defined group of organisms (say, enterotoxin-producing *Staphylococcus aureus*), there exists a spectrum of multiplication rates, metabolic activities, rates of production of toxins or other characteristic biochemical substances, sensitivities to minor variations in specimen composition, and the metabolic activity of the remaining flora. It is, therefore, futile to look for correlations between enterotoxin or any other biochemical activity (e.g. coagulase, lecithinase or DNAase) and plate counts because the aspects of the total information provided by the different methods are not the same.

Occasionally, the statistical distribution of values for each information variable happens to be sufficiently narrow that "usable" correlations are found between alternative methods and plate counts. This occurs when the commonly encountered strains of a species exhibit a narrow spectrum of the relevant metabolic activity and when the organisms are easily dispersed for the count. The correlation is fortuitous however and should not normally be expected.

It is difficult to even justify the need to search for such correlations; we are, thereby, attempting to squeeze superior data (40% of what is required) down to the limited descriptive capability of the plate count (20% or less). Nevertheless, the literature is full of attempts to do so. Relationships have been found or claimed. The usefulness perceived in them has generally changed greatly from the originating laboratory to the microbiological community among which the publications have been disseminated.

With five exceptions, milks incubated with resazurin which remained unchanged in color at the end of the hour had a standard plate count of less than 200,000 per cc.[11]

Table 1 presents the correlation coefficients for total aerobic bacterial counts and relative intensity readings of the indicator disks. . . . Reduction was better correlated to bacterial counts with methylene blue than with triphenyltetrazolium or resazurin . . . psychrotrophic bacterial counts were better correlated than total aerobic counts to reduction. . . .[12]

Unfortunately, the conventional test-tube resazurin test does not produce results that can confidently be correlated with bacterial numbers. . . . Recently, very promising studies conducted in two different laboratories . . . showed that resazurin reduction time *did* correlate with bacterial numbers in milk . . . provide statistical evidence that the resazurin reduction test can be applied to meats and that bacterial numbers can be generated. . . .[13]

This report contains the results of over 300 analyses used to obtain a correlation between resazurin reduction and bacterial numbers. . . .[14]

. . . one method . . . either pour or drop plates, yields accurate bacterial populations and the second, the PNB reduction test, offers an approximation technique, somewhat comparable in accuracy to the resazurin test in milk.[15]

There are some points in the graph which indicate that the reduction time may vary considerably in different samples having bacterial contents at the same general level. Despite this . . . this test may be used as a rough screen for the separation of samples of very high bacterial content. . . .[16]

. . . conclude that claim cannot be laid to an accurate bacterial count by the resazurin test. . . .[17]

The resazurin reduction to colorless was quite accurate for indicating the quality of meat during this experiment. . . . Procter and Greenlie (1939), Johns (1944), Straka and Stokes (1957), Ferguson et al. (1958), Wells (1959) and Walker et al. are among many workers who have correlated bacterial numbers with resazurin reduction time in poultry, meats, egg pulp, powdered eggs, milk, precooked frozen foods and fresh and frozen vegetables. In general, the relationship was closer between resazurin dye reduction time and spoilage, than between total counts and spoilage.[18]

$\Delta\%R$, "tyrosine" value and TBA number were the only quality tests which correlated closely with both bacteria count and time of storage.[19]

The agreement between impedance and plate count classification is 50, 80 and 32% for samples over 10,000 organisms/ml, between 10,000 and 1,000 organisms/ml, and below 1,000 organisms/ml, respectively.[20]

The volumes of acid have been correlated with log bacterial numbers. . . .[21]

The present experiments confirmed that there was apparently a linear relationship between extract release volume and number of microorganisms, although statistically, the correlation was not very high, particularly for beef. . . .[22]

The fact that many of these quotations draw heavily on resazurin reduction times is not accidental. Resazurin dye has thus far been to "alternate microbiologists" a godsend comparable with agar to the early enumeratists. Like agar, however, resazurin will be shown to lead to analytical data that are in no way relevant to the consumer's needs (Chapter 6) although they may have other limited uses.

The consumer would like to know whether and for how long his food will be wholesome. But all of the deterministic approaches require so much information that we are left with gaps that we can only fill statistically from our experience. Notwithstanding the possibility that some methods provide more information than others and that 40 percent of the information may be better than 20, the required extrapolation into the dark and unknown dimensions of the remaining variables is enormous. Attempting to determine food quality using anything less than all five of the required information items is rather like the scout who runs to his king with the message "Sire, the enemy number 10,000! . . ." It matters, of course, not a little to the fate of the kingdom whether the enemy is armed with pitchforks or with cruise missiles. The imprecision inherent in any analytical method failing to provide information on all five variables must be considered so great as to make its data valueless in individual situations. Such situations arise whenever, for example, decisions on release of food from factories, entry permission for imported products, and the recall or condemnation of food must be made on the basis of microbiological data. Regardless of whether or not the food conforms to a standard, it is generally impossible to be dogmatic about its true capability for causing harm or being rejected by the consumer. To be able to do so using the deterministic approach requires the gathering of too much separate information.

Existing enumeration oriented methods of analysis thus fall badly short of providing the information required for satisfactory microbiological control of food quality. The failure is to the misfortune of both consumer and manufacturer, but in the long run the consumer pays either way. When numerical standards are tight enough to minimize statistical hazards, he pays for much food unnecessarily taken out of circulation when it was not, in fact, capable of causing harm. If standards are more lax, he pays in increased risk.

THE COUNT AS A PREDICTOR OF FOOD QUALITY

To use counts as descriptors of food quality implies that we are able to extrapolate through all of the interactions between the previously mentioned variables to some future instant just before the food is eaten. Some comments by experienced microbiologists regarding the descriptive value of the count were used to introduce this chapter. However, in order to obtain some perspective on the impossibility of the extrapolation, it is useful to look at a few simple mathematical models of microbial growth in mixed cultures. The interactions can be most dramatic.

Again, to readers who dislike mathematics, I would suggest reading the text and, in particular, the graphs illustrating the equations.

A General Growth Equation

First, we need to derive a useful model of microbial growth. A simple way to do this is to regard the activity of microorganisms in a closed system as being the resultant of two competing processes. On the one hand, the rate at which the organisms tend to do anything, e.g. multiply, is proportional to their number n thus:

$$\frac{dn}{dt} = kn \qquad (4:1)$$

for multiplication, k being some constant.

On the other hand, the rate at which the organisms are *permitted* to do so is constrained by other factors, such as the supply of nutrients. As n approaches the maximum growth situation ($n \rightarrow n_\infty$), this rate must obviously fall to zero, thus:

$$k = \frac{k_0(n_\infty - n)}{n_\infty} \qquad (4:2)$$

where k_0 is the initial value of k.

The overall multiplication rate is the product of (4:1) and (4:2) and can be written

$$\frac{dn}{dt} = \frac{k_0 n(n_\infty - n)}{n_\infty} \qquad (4:3)$$

which on integration becomes

$$n = \frac{n_0 n_\infty e^{k_0 t}}{\left[n_\infty + n_0(e^{k_0 t} - 1)\right]} \qquad (4:4)$$

n_0 being the initial number of organisms, i.e. the inoculum.

Competition for Nutrients in Mixed Cultures

Equation (4:4) described the nutrient limited growth of a pure culture from inoculation to its maximum population level. Consider now the growth of a mixed culture of two species in which

for species 1, we have k_1, n_1, n_{1_0}
for species 2, we have k_2, n_2, n_{2_0}

and where, when growth is a maximum, we have

$$n_1 + n_2 = n_\infty$$

If the organisms do not interact antagonistically or synergistically in any way but only affect one another through competition for the same limiting nutrient, it is easy to show that

$$\frac{n_1}{n_{1_0}} = \left(\frac{n_2}{n_{2_0}}\right)\frac{k_1}{k_2} \tag{4:5}$$

To illustrate a practical situation, consider a volume, inoculated equally (at low level) with two species, both normally able to achieve the same maximum population density (say 10^8 orgs/ml) but where one multiplies twice as rapidly as the other ($k_1 = 2k_2$). Division times, for example, might be 15 and 30 minutes. A plate count will enumerate both species because

a. the organisms will be separated and will thus not be in competition.
b. even the slowly growing species will be able to form visible colonies, i.e. $> 10^8$ organisms, from single cells during 24 hours of incubation.

In the environment of the food, however, the organisms will be in direct competition for nutrients. In the absence of other variables to affect relative growth rates, Equation (4:5) will describe their behavior. Figure 4-1 illustrates how the slower growing species, since it manages to reach a maximum population density of only 10^4 organisms/ml, will have an insignificant effect on the deterioration of the food. Note, also, that had it been present initially at a concentration 1,000 times greater than that of the faster growing organism — a concentration at which it would certainly have been the only species detected — it would still have been outgrown by a factor of 10 within the food.

This is a very simple model, but it illustrates quite dramatically how valueless it may be to enumerate a species without also measuring multiplication rates for all species present in a food.

Neutralization Effects

Very commonly, in mixed cultures, species may interact indirectly with one another through their opposing effects on a pool of some essential substance in the food that may be inhibitory or toxic to the organisms. The same interactions may affect the course of deterioration of a food.

Food Microbiology

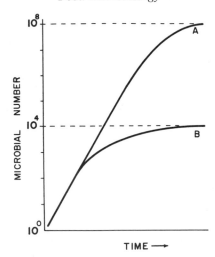

Figure 4-1. A fast-growing microbial species (curve A) may reach almost its normal maximum population density in a mixed culture. A slowly growing species, however (curve B) may reach an insignificant proportion of the total population. (Schematic illustration of Equation 4:5.)

For example, the production of ammonia (NH_3) by one species and its removal, e.g. by neutralization as NH_4^+ by another in a food for which the pH is an important parameter for organoleptic acceptability will be rather susceptible to the relative levels of these species as they multiply together. Consider in general that a metabolic product p is produced in quantity Q_p proportionally to the number of organisms of species p which have so far grown.

$$Q_p = K_p n_p \qquad (4:6)$$

where K_p is a constant. We can use this expression in Equation (4:5) to describe n_p. In just the same way, the "antiproduct" a is produced in quantity Q_a by the other species:

$$Q_a = K_a n_a \qquad (4:7)$$

and we shall also use this expression in Equation (4:5).

The net amount of p produced or removed from the pool at any instant is simply:

$$Q_p - Q_a = \frac{K_p n_{p_0} n_{p_\infty} e^{kpt}}{n_{p_\infty} + n_{p_0}(e^{kpt} - 1)} - \frac{K_a n_{a_0} n_{a_\infty} e^{kat}}{n_{a_\infty} + n_{a_0}(e^{kat} - 1)} \qquad (4:8)$$

Now for a practical example. Consider an NH_3 producer, initially present at a concentration 10,000 times greater than an H^+ producer (i.e. NH_3 remover) but like before, multiplying at only one-half its rate. If we simplify calculations by again assuming that both have maximum population levels of 10^8 organisms/ml, Equation (4:8) reduces to

$$Q_p - Q_a = \frac{10^4 e^t - e^{2t}}{10^8} \qquad (4:9)$$

where t is the number of doublings of species a. Figure 4-2 illustrates how the pH will change as the organisms multiply and interact within the food. Note how the pH may at first rise because of the preponderance of NH_3-producing (e.g. proteolytic) organisms initially present. Indeed, its behavior may not be noticeably different from another sample of the food containing no H^+ producers at all. Suddenly, however, its pH drops dramatically as the acid producers take over.

Note that at a proportion of only 1 in 10,000, the acid producers would almost certainly have been missed by a plate count analysis of the food. Even had they been detected, their significance would probably have been overlooked in view of their minute proportion. This example again illustrates how valueless numerical data may be in the absence of information on the other relevant variables.

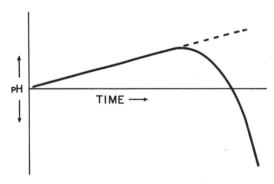

Figure 4-2. The pH of a food may swing dramatically when mixed cultures are present. Schematic illustration of the course of pH with time, derived from H^+ flux described by Equation 4:9. Dotted line indicates the smooth course pH would have taken if only NH_3 producers were present.

Reactions Between Metabolic Products

As a final example, consider a situation where metabolic products of one species react with a metabolic product of another, neutralizing it or destroying it. Or the first metabolic product may be metabolized by the second species. For example, species P might produce a toxic protein that is destroyed during the proteolytic activity of species A.

Designating the product of interest by P and the reactant by A, we can write the equation for the reaction as

$$P + A \rightarrow \text{Products}$$

where the active masses of the reactants are

$$[P] = (P_{produced} - P_{consumed})$$
$$[A] = (A_{produced} - A_{consumed})$$
$$= (A_{produced} - P_{consumed})$$

and the rate of reaction is given by

$$-\frac{dP}{dt} = K[P][A] \tag{4:10}$$

$$= K(P_{produced} - P_{consumed})(A_{produced} - P_{consumed}) \tag{4:11}$$

If the reactants are produced at rates proportional to the numbers of their respective organisms, one can arrive at the rather awkward expression:

$$-\frac{dP}{dt} = K(k_P e^{kPt} - K[P][A])(k_A e^{kAt} - K[P][A]) \tag{4:12}$$

Taking, as before, a situation where species P is initially present at 10,000 times the level of species A (insofar as product P is concerned) but only multiplies at one-half its rate, one can arrive at the following approximate solution to Equation (4:12):

$$[P]^2 - [P]\left(1 + e^t + \frac{e^{2t}}{10^4}\right) + \frac{e^{3t}}{10^4} = 0 \tag{4:13}$$

Figure 4-3 illustrates the typical shape of this curve. If the product P was a hazardous toxin, it should be noted that an analysis of the food based on numbers of organisms might wrongly condemn it as being capable of generating toxin,

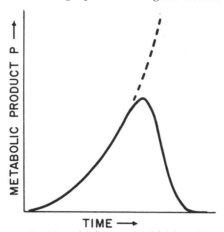

Figure 4-3. Schematic illustration of Equation 4:13, showing how in a mixed culture one microbial metabolic product may first appear normally (dotted line) then disappear as it reacts with products from, or is consumed by, the other species.

whereas, in fact, after first forming in small quantities, the toxin level would fall again and remain very low.

The Future of Microbial Counts

I leave this chapter with the following conclusions:

1. Efforts to simulate one type of microbiological data, e.g. numbers of organisms, using other approaches, e.g. metabolic activities, are unlikely to be satisfactorily rewarded.
2. The popularity that has developed for enumerative methods of assessing food microbiological quality should be considered truly remarkable in that, not only do numerical methods not provide information useful to the consumer, they rarely even provide good numerical data and, in particular, they provide the smallest quantity of relevant information out of the various possible approaches.
3. No accepted methods of analysis other than those using simulated consumers provide good microbiological information that is directly relevant to the consumer and useful for control in any way other than on a very broad statistical basis.
4. Future developments in methodology, instrumentation, or automation should not be directed towards the production of

numerical data or numerical data substitutes. It is illogical and unjustifiable to attempt to popularize methods yielding an uncertain substitute for a datum of questionable value.

REFERENCES

1. Watrous, G. H., Jr., Barnard, S. E. and Coleman, W. W., II: A survey of the actual and potential bacterial keeping quality of pasteurized milk from 50 Pennsylvania dairy plants. *J Milk Food Technol, 34:*145, 1971.

2. Winslow, R. L.: Bacterial standards for retail meats. *J Food Protect, 42:*438, 1979.

3. Hankin, L., Dilman, W. F. and Stephens, G. R.: Keeping quality of pasteurized milk for retail sale related to code date, storage temperature, and microbial counts. *J Food Protect, 40:*848, 1977.

4. Wehr, H. M.: Attitudes and policies of state governments. *Food Technol 32:*63, 1978.

5. Morrison, S. M.: Microbiology standards for waters. *J Food Protect, 41:*304, 1978.

6. Hobbs, B. C.: Problems and solutions in food microbiology. *Food Technol, 31:*90. 1977.

7. Stiles, M. E.: Reliability of selective media for recovery of staphylococci from cheese. *J Food Protect, 40:*11, 1977.

8. American Public Health Association. *Standard Methods for Analysis of Diary Products,* 12th ed. New York, A.P.H.A. Inc., 1967.

9. Fowler, J. L., Clark, W. S., Foster J. F. and Hopkins, A.: Analyst variation in doing the standard plate count as described in SMEDP. *J Food Protect, 41:*4, 1978.

10. Silliker, J. H., Gabis D. A. and May, A.: ICMSF methods studies. XI. Collaborative/comparative studies on the determination of coliforms using the most probable number procedure. *J Food Protect, 42:*638, 1979.

11. Barrett, W. D., Rutan, H. and Keenan, J. A.: The resazurin test — its use and practicability as applied to the quality control of raw milk. *J Dairy Sci, 20:*705, 1937.

12. Emswiler, B. S., Kotula, A. W., Chesnut C. M. and Young, E. P.: Dye reduction method for estimating bacterial counts in ground beef. *Appl Environ Microbiol, 31:*618, 1976.

13. Holley, R. A., Smith, S. M. and Kempton, A. G.: Rapid measurement of meat quality by resazurin reduction. I. Factors affecting test validity. *Can Inst Food Sci Technol J, 10:*153, 1977.

14. Dodsworth, P. J. and Kempton, A. G.: Rapid measurement of meat quality by resazurin reduction. II. Industrial application. *Can Inst Food Sci Technol J, 10:*158, 1977.

15. Mallmann, W. L., Dawson, L. E., Sultzer, B. M. and Wright, H. S.: Studies on microbiological methods for predicting shelf-life of dressed poultry. *Food Technol, 12:*122, 1958.

16. Procter, B. E. and Greenlie, D. G.: Redox potential indicators in quality control of foods. *Food Res, 4:*441, 1939.
17. Ramsdell, G. A., Johnson, W. T. and Evans, F. R.: Investigation of resazurin as an indicator of the sanitary condition of milk. *J Dairy Sci 18:*705, 1935.
18. Saffle, R. L., May, K. N., Hamid, H. A. and Irby, J. D.: Comparing three rapid methods of detecting spoilage in meat. *Food Technol, 15:*465, 1961.
19. Strange, R. E., Benedict, R. C. Smith, J. L. and Swift, C. E.: Evaluation of rapid tests for monitoring alterations in meat quality during storage. *J Food Protect, 40:*843, 1977.
20. Cady, P., Hardy, D. Martins, S. Dufour, S. W. and Kraeger, S. J.: Automated impedance measurements for rapid screening of milk microbial content. *J Food Protect, 41:*277, 1978.
21. Shelef, L. A. and Jay, J. M.: Use of a titrimetric method to assess the bacteriological spoilage of fresh beef. *Appl Microbiol, 19:*902, 1970.
22. Lowis, M. J.: The role of estract release volume in a rapid method for assessing the microbiological quality of pork and beef. *J Food Technol, 6:*415, 1971.

Chapter 5

A FRAMEWORK FOR THE FUTURE

IN PRECEDING CHAPTERS, I attempted to show that while the numerative approach to food microbiology has been fairly successful — in a statistical sense — for controlling and interpreting food quality, its shortcomings cost us dearly and, perhaps more seriously, inhibit the development of food microbiology as a modern, instrumented science. I believe many of these shortcomings can be eliminated but only if we take the rather drastic step of laying aside many of our accepted and trusted values in favor of a very different approach, such as the one proposed in this chapter.

To be acceptable, a new approach must meet many requirements, for many parties are interested in food microbiology from many points of view. Microbiologists, for example, would generally expect the new approach to facilitate rapid commercial development of satisfactory instruments so that it can be quickly and widely implemented. Instrument manufacturers would hope it might relieve prohibitive development costs of automated microbiological instruments and the need for elaborate interlaboratory performance studies. Food manufacturers are anxious for faster methods of microbiological analysis. Regulatory agencies would like microbiological data to be absolute or unequivocal since this would greatly facilitate enforcement of national hygiene or quality standards. Quality control microbiologists would certainly welcome the authority carried by unequivocal data, and food manufacturers themselves, in general, would probably prefer there be less room for dispute over potential hazards or losses. Lawyers ought also to prefer unequivocal analytical findings; however, since a reduced arguability of legal proceedings could pose a minute financial threat for them, they might be less enthusiastic to this idea. Consumers and consumer associations would justifiably argue that microbiological data should be demonstrably relevant to those likely to be affected rather than to a less personal national statistic. And so it goes on and on. The main requirements can thus be stated as

1. to provide standards defining acceptable levels of specific hazards rather than hazards that are implied but not quantified as at present
2. to provide data that are precise and immediately interpretable as the magnitude of the hazard
3. to provide unequivocal data so as to facilitate enforcement
4. to allow a greater utilization of existing technology and minimize the need for collaborative studies so as to facilitate commercial development of instruments
5. to provide for faster analyses

The approach that will be outlined goes some way towards meeting these needs. Any novelty in its objectives, standards, and methodologies derives from one change — its focus is people. While not denying the pertinence of microorganisms to the microbial spoilage or hazards of food, it deemphasizes their value as descriptors of quality. This viewpoint by no means implies devaluation of conventional studies of microbial taxonomy, biochemistry, and other behaviors.* It is simply an acceptance that we are not yet in sight of the day when the deterministic approach provides sufficient information for the satisfactory control of food quality. This chapter introduces the idea and indicates its attractions. More detailed aspects and practical considerations are left to Chapter 6.

VALUABLE AND LESS VALUABLE INFORMATION

By far the most important aspect of microbial contamination in food is whether or not it causes undesirable psychological or physiological responses in humans when it is inspected for purchase or eaten. No other aspects, such as the number of bacteria

---

* Indeed, such studies are essential if — in a distant future — instruments able to rapidly *predict* the growth of unwholesome parameters in a food are ever to be developed. It is conceivable, for example, that instruments might be invented capable of rapidly analyzing DNA and other macromolecules contained in a food, and from the data, the probable sequence of biochemical events over any required time/conditions span could be computed.

contained in one gram, their staining characteristics, species, biotype, serotype, optimum growth temperature, the personal hygiene of plant personnel, speed of distribution, storage conditions, etc., are as immediately important, provided the consumer is not aware of them or is not prejudiced by influences from his upbringing, his friends, journalists, advertising agencies, and so on.

That is by no means to say that these other aspects are either irrelevant or uninteresting. They are invaluable indicators of the general excellency (or otherwise) of manufacturing practices or the sources of epidemics, for example. But the quality of the individual specimen he consumes is basically more important to the consumer. Other things being equal, it does not matter to the consumer's body whether he eats a microbiologically superior sample from a disreputable factory or an average sample from a better one, provided he is not told and provided he does not feel sick afterwards.

If there is any reason to criticize the conventional numerical approach to microbiological analysis, it is in its tendency to emphasize the importance of microorganisms rather than the consumer, for no person ever wrinkled his nose at or became ill from microorganisms themselves (unless by thinking about them). Human bodies are quite incapable of detecting microorganisms; our unaided senses do not have the necessary sensitivity and selectivity. We respond only to the accumulated physical, chemical, biochemical, and immunological manifestations of microorganisms and *then* only when these parameters have reached sufficiently high levels.

It is not at all pedantic to emphasize that we do not pass over a stale pack of sliced meat in favor of another one at the supermarket because we see bacteria on the surface but rather because of color and reflectance changes resulting from their metabolism, and then only when these have progressed far enough for us to be able to recognize an undesirable hue. We throw out ground beef not because we smell bacteria but only hydrogen sulfide, ammonia, or fatty acids resulting from their metabolism. We suffer fevers and diarrhea not because of the microorganisms multiplying within us but because our bodily chemistry is perturbed by toxins, enzymes, antigens, and other products related to their metabolism. The cumulative effect is what matters. It is certainly essential for the appropriate microorganisms to be there, but their presence and number alone mean very little to the wholesomeness of a food. Multiplication rates, the magnitude of the relevant metabolic activities, and the interactions occurring between components of the microbial flora in the environment provided by the food, together with human thresholds of perception or response are equally important. These do not come into routine, analytical objectives in the conventional enumerative approach to food microbiology. At best, they are accounted for probabilistically through experience (Chapter 4).

To the consumer, this loss of information may be rather significant. From his point of view, the most valuable measurement we could make on a food is not the number or species of organisms in it but their potential for generating in it (before it is eaten) undesirable changes at levels to which his body is likely to respond. We can go a long way towards making such measurements.

PARAMETERS OF UNWHOLESOMENESS

The term *parameter of unwholesomeness,* introduced in Chapter 4, will be used frequently from now on. It describes any undesirable physical, chemical, biochemical, immunological, or other property of a food that is capable — sooner or later — of being detected by (i.e. causing a response in) the human body when it reaches a suitable level or *amplitude.* It is not quite the same as the term *stimulus,* which is widely used in sensory evaluation and other related disciplines. A parameter only becomes a stimulus when it exceeds the consumer's physiological or psychological threshold.

Typical parameters of unwholesomeness are color; hue; reflectance; tensile strength; elasticity; viscosity and other rheometric properties; phase separations; partial pressures of mercaptans, ammonia, amines, fatty acids, ketones and other odorous substances; concentrations of hydrogen ions, glycosides, peptides, purines and other taste-stimulating substances; concentrations of toxins; antigenic sites; and so on. For strictly microbiological purposes, a parameter of unwholesomeness derives from microbial presence or metabolism, though in a wider sense this need not be so. Enzymic activities inherited from the original tissue, for example, may cause undesirable changes that are not easily distinguished from those due to microbial activity. In practice, these intrinsic activities are just as important as microbial ones; they ought to — and can easily be — taken into account in describing the microbiological quality of a food.

Defining these parameters accurately, particularly the organoleptically significant ones, cannot always be expected to be simple, of course. Our senses are sometimes exquisitely perceptive yet at other times remarkably fallible. The exact physical or mechanical parameters affecting tenderness, crispness, or chewiness, for example, are very difficult to pin down. Similarly, so many chemicals contribute to our perception of a single odor or flavor that we cannot expect to be able to define each one as an important parameter to be measured. However, the food industry has already developed satisfactory methods of measuring many of the most important organoleptically significant parameters, and those that have not already been identified are certainly not beyond reach. The toxins and substances of immediate immunological significance such as allergens would seem to be relatively easy to define since their important properties are much more specific. Thus, while existing methodologies for determining staphylococcal enterotoxins, for example, are somewhat messy, there would appear to be no reason why analytical methods suitable for their instrumental determination cannot be developed. Defining the relevant parameters for infectious agents such as the *Salmonella* bacteria or enteroviruses may be somewhat more of a problem, although not insoluble. The case of infectious agents is discussed again later (Chapter 6).

A parameter of unwholesomeness may be regarded as a signal that the body — as a detector with rather unstable characteristics — detects against a background of random fluctuations or noise (see Chapter 3). The amplitude attained by a parameter at the instant before consumption or inspection for purchase determincs (as will be described later) its probability of being detected and causing a response. One cannot escape the effect of the relative strengths of signals and noise in detectability. This is why our bodies appear to exhibit thresholds — both in the perception and acceptance of all the possible parameters.

A threshold may be defined here as the amplitude of a parameter needed to cause a physical, psychological, or physiological response. It may be, for example, the odor level over a food at which a would-be consumer is sufficiently strongly motivated to seek another store or to throw the food in the garbage. Or it may be a level of toxin at which the unwitting consumer subsequently falls ill. This is slightly different from threshold in the normal context of sensory evaluation. The American Society for Testing and Materials,[1] for example, defines six thresholds for odor perception (detection, difference, recognition, supra, absolute, and terminal) that are determined under carefully controlled and rather artificial conditions. None of these can be expected to relate simply to the consumer's reactions on, say, opening a pack of luncheon meat. This depends on more complex factors than an absolute odor detection threshold; however, the experimental requirements for determining thresholds of response will often be simpler than those required for conventional sensory evaluation.

A population of humans will exhibit a statistical distribution of responses to a given parameter since thresholds vary not only from person to person but with his/her health, daily cycle, mood, previous exposure to the parameter, education, the context of the food, and possibly with many other factors. Our threshold of response in rejecting food on account of its partial pressure of propionic acid vapor, for example, is likely to be much higher when the subject is a Camembert cheese than when it is a strawberry parfait. Similarly, we are more likely to respond with a show of disgust to slime on a weiner during the Christmas holiday (our threshold will be lower) than during a backpacking holiday. Likewise, a North American executive might (on the grounds of its odor of hydrogen sulfide) reject meat that would be perfectly acceptable to a Guinea tribesman,* his threshold of response to the relevant parameter (partial pressure of hydro-

---

* This is conjectural. I have no knowledge of the true relative acceptabilities of H_2S odor above meat for these different ethnic groups. If either party is offended, I apologize.

gen sulfide) being lower. And a population normally exposed to food of poor microbiological quality may have a much higher average threshold to, say, staphylococcal intoxication than one more fortunate. In part, this may result from a lowered signal to noise ratio, caused by interference from numerous other food-borne sources of discomfort. It may also result from an accumulated immunity to the toxins.

AMPLITUDE/RESPONSE CURVES, AND HOW VERY VALUABLE INFORMATION MIGHT BE GENERATED

Because few humans are identical, and even an individual varies physiologically and psychologically from moment to moment, any population of humans will exhibit a spectrum of apparent thresholds of response to a given parameter of unwholesomeness in food. Figure 5-1 illustrates some types of spectra that might be expected. Consider a hypothetical experiment on the parameter represented by the partial pressure of hydrogen sulfide vapor over ground beef. (This chemical contributes significantly to the unwholesome odor of ground beef that has incubated too long in an airtight pack, although its characteristic odor is not necessarily recognizable. It can be detected by, for example, its reaction with lead acetate held in the headspace; also, ground beef incubated similarly, but in the presence of lead acetate, has a much more wholesome odor.) Packs of ground beef containing various levels of sulfide are prepared, the pH adjusted to release the vapor, and a panel of

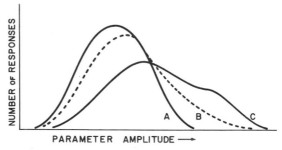

Figure 5-1. Possible types of response spectra to parameters of unwholesomeness. A — the population exhibits a statistically normal distribution (unusual); B — the population exhibits a skewed distribution; C — the population contains two phenotypes of different sensitivities.

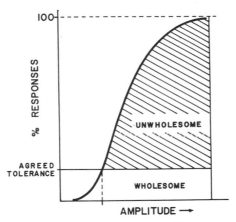

Figure 5-2. Schematic amplitude/response curve for a parameter of unwhole-someness, such as might be drawn for expressions of disgust evoked by H_2S odor from ground beef. Interested parties may agree on the percentage of responses tolerable from a food; a parameter amplitude at the threshold of unwholesomeness can be derived from the amplitude/response curve.

subjects asked to open the packs under conditions simulating preparation of supper. Panelists are asked to report whether they would continue preparing the meal for their family or whether the odor from each pack would be likely to make them discard the meat.

If the panel was large enough, a curve rather like Figure 5-2 might be drawn from the data. If the panel had been chosen to be suitably representative of ground beef users, the curve would become a base from which very useful and unequivocal data could subsequently be derived.

For example, a committee of plant managers, marketing, public relations, financial, and other staff in a large food company, might agree on the maximum tolerable percentage of sales resulting in complaints of rotten odors from the company's ground beef patty line. The agreement would weigh immediate sales losses (decreased profitability) and estimated long-term losses from damage to the brand-image against increased manufacturing costs incurred in producing a better quality product.

Once this is decided, the company's quality control microbiologist can be presented with a singularly meaningful objective — to measure the hydrogen sulfide generating capability of patties (in terms of partial pressures or likely percentage of complaints) and to ensure that patties capable of exceeding the agreed maximum acceptable level before their printed expiry data are not released for sale. Here, therefore, the availability of an amplitude/response curve allows a patty microbiological standard to be defined in very specific and absolute terms. The standard does not mention microorganisms — it merely places a limit on their collective hydrogen sulfide generating capability. Obviously, the laboratory procedures required to enforce this standard are going to be somewhat unconventional.

To measure patty quality absolutely, using the conventional deterministic microbiological approach would be impossible, or at least, impractical. The microbiologist would need to obtain analytical data, not only on the numbers of the different organisms capable of producing hydrogen sulfide at the instant of sampling, but also on their individual rates of producing the gas during the (constantly changing) conditions they experience in the patty, their (constantly changing) multiplication rates, and the effect (also constantly changing) of other organisms on this total hydrogen sulfide producing capacity. He would need a computer to combine or integrate all this data into a vapor pressure prediction and, unless he had an amplitude/response curve similar to the one described, he would need to accumulate data relating complaints to predicted odor levels for many months before he could use his deterministic approach effectively. However, if he abandons his natural inclination to enumerate microorganisms, he will find that all he needs to control this odor parameter is an incubator, a timer, and some simple instrumentation, e.g. a colorimeter, allowing him to measure hydrogen sulfide in the headspace of patties. His experimental data will automatically be the mathematical combination of all of the above variables.

In Chapter 1, I described the dilemma of a microbiologist faced with an individual rather than overall situation. If a batch of patties, for example, is just over the limit he has set through experience, should he or should he not release that batch for sale? And how can he justify his stand against pressure from a plant manager committed to shifting products out for sale?

Conventionally, he has difficulty because of the inadequacy of the information, with only experience as a not too reliable guide. By measuring the actual organoleptically significant parameters (such as hydrogen sulfide) against the appropriate amplitude/response curve, however, he can present much more informative data about the patty quality. For example, "23 percent of the patties unsold after seven days are likely to result in complaints of rotten odor." This is a precise, direct, and unequivocal description of patty quality. To the plant manager it will be far more valuable, far more convincing, than the alternative statement, "the standard plate count is $6 \times 10^6/g$" (implying — "In my opinion, this could mean problems for you. . . ").

This example ignored several important practical details; there is obviously much more yet to discuss. However, it should be apparent that describing food quality in terms of the observable consequences of the microbial contamination rather than in conventional terms of microbial numbers has some attractions. In fact, the example already suggests that a good proportion of the requirements of a new approach (listed earlier) might be neatly fulfilled. Thus, the hydrogen sulfide standard defines an acceptable level of a specific complaint (not a hazard, in this case), the data are unequivocal, they are immediately interpretable as the magnitude of the complaint, and an existing instrument technology has been used so that little development is necessary.

Amplitude/response curves like Figure 5-2 can, in principle, be drawn up for any parameter of unwholesomeness. In toxicology, they are known as dose/response curves. Data are generated by established methods; the objective is usually to determine LD_{50} values or (in microbiology) minimum infective doses or ID_{50} values. Similar curves can occur in sensory evaluation. However, it is not essential to determine the whole curve. If an acceptable percentage of responses to a parameter has been previously agreed upon, only the lower portion of the curve, i.e. from zero response up to the agreed level, need be obtained. This is important when the parameter is a hazardous one, such as a toxin or disease-producing organism, if data for the curves are generated by people. It is obviously much easier to produce amplitude/response curves for organoleptically significant parameters than for toxic or infective ones. However, the latter need not be completely inaccessible, and it is not absolutely necessary that the data result from deliberate experiments.

In the previous patty example, food quality was assessed — in terms of one parameter only — by actually measuring its development under simulated use/abuse conditions. It would be difficult to imagine a more straightforward approach; there is no element of extrapolation or prediction involved. However, like the conventional counting approach, it involves a rather lengthy incubation, and it may not be too valuable when the food is perishable if results must be related to the actual lot from which samples were drawn. Such direct and unequivocal methods of obtaining data might be most attractive to regulatory and other agencies who are concerned with the disciplinary value of compliance analyses rather than with spoilage. If the quality control microbiologist wished to predict the shelf lives of individual patty lots, he would be advised to use an analytical method encroaching less on the patty lifetime.

In order to provide early keepability information, data must be predictive, i.e. they must allow extrapolation to shelf intervals outside the experimental range. The parameter of unwholesomeness approach allows predictive measurements to be made. For example, analytical data from identical samples incubated under different conditions can be both interpolated and extrapolated. As with all predictions, a measure of uncertainty is then added. However, it can be argued (Keepability Predictions by Extrapolation, Chapter 6) that this uncertainty is much smaller than if time-to-spoilage predictions were to be made simply from microbial counts.

This approach to assessing food quality might be described as *stochastic* — in contrast to the determinism implied in our conventional approach. Accepting it begins with acknowledging our inability to determine all those interacting variables lying between the conventional analytical answer (microbial number) and the one we would like (potential hazard or complaint). Instead, we set up more easily handled schemes that are closer to the desired answer. If we sacrifice information about microbial number, we can gain considerably in confidence about a more valuable datum — how the food may affect us. In the right situation, this answer can be very precise, although the exact interplay of variables by means of which it was reached may be open to conjecture. Compare this with the conventional numerical approach in which we may be fairly sure about our measurement of one variable but find the really useful answer is open to conjecture.

In the long run, data about the true parameters of unwholesomeness will be just as valuable as microbial counts for monitoring, controlling, or enforcing good manufacturing practices. If the microbiological quality of food could be controlled completely at the manufacturing stage, equality alone might provide little incentive for changing our approach. Indeed, there might be little need for microbiological analyses at all. We do not live in such an ideal world, however. For many years to come, decisions to permit the distribution and consumption of foods will have to be made on the basis of microbiological analyses. If the data can be obtained directly in terms of the true hazards or complaints rather than microbial numbers that relate only probabilistically to them, I believe there is justification for moving that way.

MICROBIOLOGICAL STANDARDS FOR FOODS

Deterioration of food quality as a result of microbial activity is a continuing process, terminated only when the food is consumed. A microbiological standard must, therefore, relate to a particular instant in the life of a food. Conventionally, this is the instant of sampling and cannot be otherwise, for the microbial count is a destructive analysis. It will be obvious, however, that the concept of the maximum acceptable amplitude of a parameter of unwholesomeness relates much more closely to the instant of consumption (or inspection for purchase) of a food.

In Chapter 6, it will be shown how it is possible (in principle, at least) to develop instruments capable of providing a veritable envelope of data describing the microbiological quality of a food after various storage times and conditions. We do not need to consider such magnificent sophistication here; it is merely necessary for us to place reasonable limits on the incubation interval and conditions before which a food should be expected to be consumed. If a little extra research is carried out to determine (more or less) the optimum generation conditions for a particular parameter, a standard to cover the worst possible case of abuse may even be defined. As an example, consider how a standard for staphylococcal enterotoxin B in ham might be formulated.

1. Using, perhaps, the accumulated information from outbreaks of staphylococcal intoxication from ham (manufacturing and distribution practices, domestic preparation and storage conditions, likely doses of toxins consumed, etc.) as a starting point,

2. representatives from the various interested groups (food industry, microbiological societies, regulatory agencies and other health oriented administrations, consumer's associations, sociologists, etc.) tentatively agree a stress condition, e.g. time/temperature abuse, ham can reasonably be expected to suffer before consumption and

3. an acceptable limit (0.1% say) to the percentage of the population liable to suffer staphylococcal intoxication if they are unwise enough to so abuse ham before eating it.

4. Sufficient data to enable the dose/response curve for staphylococcal enterotoxin B to be drawn as far as the agreed 0.1 percent response level are now collected (see Methods for Generating Amplitude/Response Curves, Chapter 6).

5. The most probable dose of toxin causing 0.1 percent response in the population is calculated, and

6. the committee reconvenes. If all goes well, the experimental findings are accepted, the agreed response level is affirmed, and a proposal for a standard forwarded through the appropriate ministry to eventually become a statutory requirement for the microbiological quality of ham.

The standard may be worded so as to make it clear that, for example:

". . . the level of toxin in a 100 g sample shall not exceed X μg in less than 24 hours at 25°C . . ." or, ". . . 100 g samples shall not be capable of generating toxin at a level causing responses in more than 0.1 percent of the population in less than 24 hours at 25°C. . . ." A statute such as this does not attempt to control microbial content directly, although the manufacturer will not miss the statistical implication of microbial numbers. Instead, it shifts the essence of violation and its consequences to a scientifically demonstrable ability of the food to cause harm. That is to say, it regulates the *performance* of microorganisms in a food instead of their mere occurrence. The quality description defined by the standard (and, therefore, any microbial analyses we might make) has effectively been shifted from point of distribution or sale to *the last reasonable instant of consumption*. Thus, it eliminates the terrible uncertainty we currently experience when we try to interpret analytical data in terms of actual hazard or spoilage potential.

It should immediately be obvious that the approach dictated by a standard formulated this way leads to much more valuable information where individual cases are concerned. By conventional methodology, for example, the relative potential toxin-producing ability of a food is suggested from counts of *S. aureus* colonies. Affirmation of enterotoxigenic activity usually rests on qualitative demonstration (presence, absence) of coagulase activity which, in turn, has qualitatively been shown to correlate with enterotoxigenic activity in liquid growth media. Because the potential for producing toxins is implied but not quantified, it is quite possible to condemn a food on the grounds of its coagulase-positive staphylococcus counts when, in fact, the organisms were only weakly capable of producing toxins. It is not currently feasible to assess the enterotoxigenic activity of each colony found on the agar plates even if the organisms could be shown to have the same activity in the food. A low overall limit — designed to exclude most of the bad cases — must be enforced so as to obtain an acceptably low level of incidences of food poisoning. But, other things being equal, the consumer should be able to tolerate a food containing higher counts of *S. aureus,* provided the organisms are only weakly toxigenic, and vice versa. The existence of a standard encouraging analysis of food for its actual ability to generate physiologically active toxin concentrations (or doses) would allow quality control at the national level at least as effectively as those we have at present. It would, however, allow an individual specimen of food to be assessed on its own merit. Thus, a food containing only weakly toxigenic organisms, even though they be present in relatively high concentrations, would prove itself to be of low hazard in this particular context. Similarly, a food containing an acceptably low concentration of organisms by a conventional standard might prove itself to be relatively hazardous if those organisms were of unusually high enterotoxigenic activity.

ABSOLUTE STANDARDS

So far, we have considered a standard formulated to protect consumers from the consequences of a period of reasonable abuse of a food. If, however, we can define conditions for the most rapid generation of a parameter of unwholesomeness, we are in a position to describe the quality of a food in absolute terms.

Staphylococcal enterotoxins, for example, are not normally produced most rapidly at 20°C but at a somewhat higher temperature. Thus, there is evidence that the optimum temperature for production of enterotoxins B and C in liquid growth media is about 40°C.[2] Whether this is the optimum temperature for their production in, say, ham, by the majority of staphylococci has probably not been determined, but obtaining the information requires only a straightforward experiment. The presence or absence of oxygen, as determined by the packaging of a ham, might also be similarly investigated. Temperature has a particularly important role, but other variables must be considered in so far as they are relevant to each food.

Conditions for an optimum rate of increase of a parameter may, in principle, be written into the standard, for example: ". . . the level of toxin in a 100 g sample shall not exceed X μg in less than 24 hours at 40°C in the absence of air . . . etc." so that we obtain an absolute standard. Food shown to conform to this standard could thus be certified as being incapable of producing an unacceptable level of the specific hazard, regardless of storage conditions, within the specified time. Whether this would be necessary or even desirable for the majority of foods is questionable; however, the potential for writing absolute standards exists with this approach to microbiological quality. Certainly, there are a few situations (wilderness vacationing, military operations, etc.) where such a guarantee of quality might be attractive.

In the conventional approach to food microbiology, only *one* absolute standard can be defined. This is the condition of sterility — the limiting case of microbial number — since we are sufficiently sure of our knowledge of the behavior of microorganisms to be confident that, say, drastic heat treatment of a food will reduce viable organisms to an impossibly low proportion. Any proof of the presence of viable microorganisms represents unequivocal evidence of violation when the standard is sterility. If the organisms happen to be *Clostridium botulinum* in cans, we may be rather worried about the prognosis (that someone will consume a physiologically active dose of botulinal toxin). However, when we try to move from microbes to man, stretching the conclusion to that of unequivocally demonstrated hazard is not so easy. The organism can exist harmlessly in many foods as, for example, when its growth or toxin production is inhibited by adverse pH or nitrite concentrations; likewise, potentially lethal levels of toxins may be deactivated by some ubiquitous cooking procedures. The deterministic approach must either fall back on the inefficient safety first rule that any demonstrated presence is unacceptable, or it must provide additional data such as curves showing doubling times, toxin production, etc. against pH, temperature, oxygen partial pressure, and so on so as to allow reasonably safe limits to be defined for a food. Within these limits, a few dangerous organisms escaping the processing treatments are unlikely to present hazards. Such predictions are generally checked out by determining toxin production on experimental foods, made up so that their compositions lie in and around these limits. This procedure of actually determining generation of the relevant parameter is, it should be noted, very close to the approach advocated in this book.

SEQUENTIAL MEASUREMENT OF AMPLITUDE

Having reached this point developing an argument, two ways of continuing seemed open. On the one hand it seemed reasonable to begin discussing attractions that might so far be raised for

Food Microbiology

it to provide some persuasion for reading further. On the other, it seemed most efficient to first discuss the very relevant aspect of sequential measurement, leaving all advantages to be combined in one section. The latter won. Therefore, the discussion may appear to introduce complications while justifications for even the groundwork are still somewhat nebulous.

The parameters of unwholesomeness of a food will not normally remain constant. Even if conditions are such that microorganisms are unable to multiply, changes resulting from their metabolism will accumulate. Sometimes, very similar changes will be caused by intrinsic processes in the food. Except for those manufactured with the help of microorganisms, we generally expect foods to deteriorate with age, i.e. for undesirable changes to become more obvious.

If foods behaved like simple model systems such as broth or agar monocultures, growth of the parameters of unwholesomeness by which our bodies recognize microbial contamination might be easily predicted. They do not because most foods are unstable, multiphasic structures, and their microbial contaminants constitute a diverse and unevenly distributed flora. Thus, contribution to the growth of a parameter of unwholesomeness by one strain of an organism in a food may begin at once, or it may be delayed until suitable conditions (concentrations of oxygen, hydrogen ions, oligopeptides, nucleotides, etc.) have been generated around the organism by the activity of others or by intrinsic processes in the food. Conversely, there may be competitive processes such as inhibition, neutralization or metabolism of one parameter by other components of the flora.

Therefore, while the generation of many parameters of unwholesomeness may progress relatively smoothly, they may at times increase very erratically and may even decrease at times in a manner we cannot predict simply from a knowledge of microbial numbers. It would be preferable if we could exclude the possibility of the amplitude of a parameter exceeding an agreed level of acceptability at some stage in the life of a food but decreasing to an acceptable level before it is tested for compliance with the standard (Fig. 5-3).

If we make sequential measurements of a parameter's amplitude, or — preferably — if an instrument carries out the

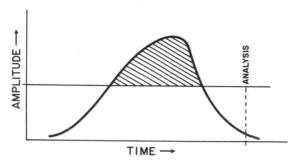

Figure 5-3. Analysis methods should prevent situations where a parameter amplitude rises to an unacceptable value but falls again before the measurement is made (compare with Figs. 4-2 and 4-3).

sequential measurements for us (Fig. 5-4), we can obviously eliminate this possibility. Suppose we then ignore the incubation interval dictated by the standard (as in ". . . shall not exceed $X\mu g$ in a 100 g sample *in less than 24 hours* at 20°C . . ."). The standard now merely defines an acceptable level of human responses or unwholesomeness. If we sequentially measure the amplitude of a parameter, we can determine the incubation interval I required before the food becomes unwholesome according to the agreed definition of unwholesomeness.

We are then in a position to make statements about the food, such as, "The food will generate an unacceptable level of staphylococcal enterotoxins in 17 hours at 20°C," or if our incubation conditions are chosen to be most favorable to the generation of a parameter, "The food will be wholesome for a minimum of 17 hours."

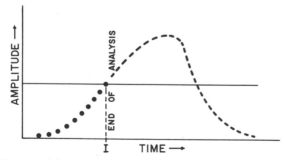

Figure 5-4. Sequential measurement of parameter amplitude prevents false acceptance and provides quantitative data for food microbiological quality.

These are very interesting and useful statements about the food's quality. The data would be far more informative to the consumer than either a microbial count or a certification of compliance (or otherwise) with a numerical standard since he can relate neither of these to the true hazard the food presents him. The same type of data would also be most useful to the industrial microbiologist. It can provide very accurate measures of product quality rather than its mere compliance with a standard. But it may be used in other ways, such as in directly identifying the contribution of any stage in a manufacturing process to overall product quality. For example, the ratio $\frac{I_2}{I_1}$ of the incubation intervals needed for samples taken on either side of a manufacturing step describes the effect of that step on product keepability. It may say, for example, that a comminuting step affects product keepability by 0.8, i.e. reduces the overall keepability of the product by 20 percent.

This is a very meaningful statement about this particular comminuting step. It could not be made through a knowledge of microbial counts before and after because too many variables must be taken into account in the resulting calculation. For example, improved dispersal of the microorganisms alone may have caused the change, whereas their number may have remained constant or even decreased under the influence of the mechanical stresses encountered. In principle, therefore, describing product quality in terms of parameters of unwholesomeness permits analyses that sum all the variables for individual situations like this one in a manner that is impossible with numerical data.

Of course, grinders such as this are very common in food factories, and they are frequently a center for microbiological attention. Training and experience would currently allow microbiologists to make very educated guesses at such a grinder's influence but, as has been stressed many times, although such experience works well statistically, it is less useful for individual situations. Thus, although comminuting steps, in general, may be known to shorten product shelf lives, the effect of any particular grinder — being a complex function of design, machining tolerances, wear, maintenance, and operating cycle — cannot be so reliably stated.

One might say that the method of Sequential Measurement until Amplitude Reaches Threshold would be a SMART way to analyze foods.

THE PROSPECTIVE MICROBIOLOGICAL ANALYSIS

It should be apparent from the argument so far that microbiological analyses based on this approach will use any parameters considered relevant to human perceptions of food quality. The approach encompasses all existing alternative and conventional microbiological analytical methods, together with others that may not yet have been applied, as part of its overall methodology. Thus, if surface viscosity and pH are considered the most significant organoleptic factors in the keeping quality of weiners, the analyses and/or instrumentation will be based on these parameters. In other situations, analyses may be based on both organoleptic and toxic or immunologically active parameters. Generally, the required analyses and limits of detection will be those for which instruments and techniques are already available (pH meters, viscometers, spectrophotometers, chromatographs, scintillation counters, etc.). In some cases, however, the numbers of microbes ingested may, for a long time, remain the only factor satisfactorily indicating their potential pathogenicity. In such cases, these are the parameters of unwholesomeness, and analyses would have to be based on very conventional counts of the organisms; however, the count itself will still generally not be the actual datum of quality.*

The overall technique and apparatus for carrying out microbiological analyses using this approach consists of

a. assembling together a suitable incubator, timer, comparators, and standards for the various parameters of interest.
b. comparing parameter amplitudes in the specimens against those of the standards, *once* after the specified time for regu-

---

* We could be so unsuccessful at describing the physiologically active features of an organism that we are forced to retain microbial number as the effective parameter of unwholesomeness. This does not mean that the analytical approach will be identical with the conventional compliance analysis since the count should strictly be made on the specimen after incubation rather than at the instant it arrives in the laboratory. Unless, that is (as it may be for dried foods), the specified incubation interval is zero.

latory work, or *sequentially* at appropriate intervals until amplitudes equate, for most other types of work.

Preparing specimens for analysis will generally be a simple affair because the first part of the analysis consists merely of incubating the food under standard conditions. However, while it may be reasonable to place a pack of sausages wholesale in the incubator, individual ingredients (particularly dry ones such as spices and thickeners) will require more elaborate treatment (see Dry Foods and Ingredients, Chapter 6).

If the analyses are to be manual "once for all" determinations of compliance, quite conventional timers and incubators will be required since the parameters of interest will merely be measured after the appropriate incubation intervals. Carrying out analyses automatically will obviously require the development of a more sophisticated incubator having some means for automatically indexing and/or permitting the detectors access to samples. From this, it is but a small, additional development step to the sequential sampling incubator capable of providing data similar to those shown in Figure 5-4.

Analytical methods used for each parameter are obviously chosen to provide a suitable limit of detection, speed, convenience, and cost. However, the actual measured amplitudes of parameters of unwholesomeness have little value, save for curiosity's sake. Only their relation to agreed thresholds of human responses (as defined by amplitude/response curves) are important. Provided reference standards can be made available for comparison; instruments and methods need not be calibrated or quantified. They need only be comparators, indicating whether parameters clear or exceed the agreed standards.

In some situations, such as in determining the keepability of a food, the incubation intervals required for specimens to generate standard parameter amplitudes (the agreed thresholds) represent the quality data. In regulatory work where yes/no answers on compliance may often be all that is required, results of single comparisons after the specified incubation intervals may be adequate. When several different parameters are measured for each specimen, violation on any one may be sufficient cause for action.

SOME THOUGHTS ON THE PARAMETER OF UNWHOLESOMENESS APPROACH TO FOOD MICROBIOLOGY

The remainder of this chapter discusses attractions or advantages of the parameter of unwholesomeness approach to food microbiology. Some details, so far glossed over for the sake of maintaining some flow of argument, are described in Chapter 6. While reading, try to keep in mind the gloomy circle of scientific and commercial problems described in Chapter 2 that have thus far inhibited automation developments in this area, and the fact that the commercial attractiveness of an approach — whether we like it or not — is vital to its success. That is to say, instruments and automatons will not find their way into the majority of food microbiology laboratories unless commercial instrument manufacturers perceive in it a sufficiently rewarding and assured area for development investment. If a new approach to food microbiology brightens its commercial image, we may yet see instrumented, automated laboratories, a wealth of precise, pertinent data, and fewer costly breakdowns in the microbiological control of food quality.

As concerns over food quality grow during the next few years, the workloads of microbiological laboratories will increase to the extent that they can probably be met only through significant automation of the analyses. Unless we provide an avenue for the introduction of the necessary instruments, food microbiologists will become a busy and harassed people.

Limits of Detection

Detecting microorganisms at conventional regulatory or guideline levels poses extreme (and generally impossible) detection limit requirements for all nonmultiplication based techniques and instruments. The use of incubation/multiplication periods in an attempt to improve signal to noise ratios introduces increasing uncertainties about the relation between analytical results and the microbial numbers generated from them (see The General Nature of Currently Practical Methods, Chapter 3). In trying to relate contamination to the instant the sample arrives in the laboratory — when levels are at their lowest — we are making the assessment of microorganisms or their products as difficult as it possibly can be.

In contrast, looking for parameters of unwholesomeness just at levels pertinent to human physiological responses allows us to take the utmost profit from signal to noise improvement *without a corresponding loss of credibility.* Detecting most of these parameters instrumentally, once they have multiplied to levels at which a significant percentage of human bodies would also respond, presents very little problem. This applies to virtually all organoleptically significant parameters, many toxins, and (perhaps more questionably at this moment) to many infection-causing organisms. In many situations, we can expect thousand- or millionfold differences in levels between what is conventionally permitted (and, therefore, what is required to be detected) and those that cause responses in humans.

Relaxing required limits of detection by such large factors can so greatly reduce analytical problems as to open the choice between a multitude of simpler, previously impractical techniques. It is almost axiomatic that the less demanding we make an analytical requirement, the less expensive will be the commercial development of instruments to satisfy it. If suitable instruments already exist, we might at least expect to see less expensive ones in future. At best, however, suitable instruments may be developed where none previously existed.

Available Technology

We do not have complete instruments for automatically enumerating microorganisms in foods because the problems of developing them are so severe. It is, moreover, unlikely that they will be developed in the foreseeable future (Chapter 2). However, it will be very evident that shifting our analytical requirement from the unique and difficult enumeration of microbial clusters (see Indeterminacy, Chapter 4) to the measurement of more common and much more basic physical, chemical, biochemical, or immunological parameters would place at our disposal a vast array of instrument and analytical technology. Regardless of whether instrument technology was developed for these areas because the market was larger, financially more favored, or more glamorous, most of the technology required to carry out food microbiology this way is already available.

The simplest usage level for the parameter of unwholesomeness approach is in compliance analysis (one-shot analyses). Here, it could certainly be introduced with little further ado than acquiring analytical instruments for the chosen parameters. Its use in analyses related to keepability would, however, require development of the *sequential sampling incubator,* that is, an incubator (or incubators) allowing repeated access to samples. This is by no means an insurmountable technical obstacle; indeed, even if such incubators do not already exist, the engineering problems in developing them are certainly minute compared with those involved in mechanizing enumerative microbiology.

For much of food microbiology, therefore, instrument companies could construct analytical systems directly from — or after very little modification of — their existing instrument lines. Prospects of greatly reduced costs for development and testing, reduced problems or complaints resulting from handling unknown technologies, and more efficient utilization of existing design, manufacturing, distribution, servicing, stock control and other facilities could combine to make venturing into the marketing of such instruments infinitely more attractive to these companies than it is ever likely to be for conventional microbiology. The likelihood of us ever seeing suitable instruments for food microbiology is, therefore, much greater than if we cling to concepts of enumeration of microorganisms.

Comparator Simplicity

For any parameter of unwholesomeness, only one amplitude (the agreed threshold) has great significance. While this may be derived from absolute measurements, it is not essential to make absolute measurements of parameter amplitudes during the microbiological analysis of a food specimen. Indeed, in practice, it may not be necessary for a detector to display values at all; it will be sufficient if it merely indicates reliably whether or not the specimen exceeds this threshold value. That is, it need only make comparisons. Provided suitable internal standards or controls are available, precise calibration becomes unimportant. Thus, although many calibratable instruments and analytical techniques are available for immediate use, future developments for food microbiology might often profit from the inherent simplicity of comparator design. The attractions of comparators are seen in such items as reduced design, engineering, stability, and range problems.

Holistic Data

As it develops in an incubated food, a parameter of unwholesomeness combines and integrates all of the possible interacting variables in a manner currently quite impossible through mere mathematical manipulation of plate count data. However, because specimens are incubated under typical storage conditions, compliance and other analyses based on the stochastic approach would actually determine the reality.

Directness and Unequivocality

The significance of microbial standards describing maximum levels of specific hazards or complaints tolerated from foods should be easily understood by most people. This applies to the average citizen as well as to factory management, politicians, lawyers, and others who might be concerned at some time with food quality. They might not immediately have greater impact on or be more acceptable to the average person than microbial numbers since the image of "numbers of germs" has great psychological significance for us — as long as we do not have to think too closely about its true meaning (see Chapter 7). However, anyone forced to think about the logical consequence of microbial contamination, e.g. "What does this mean to *me*?" would quickly find the meaningfulness of this type of description asserting itself. The standard and the description are very direct.

The comment was made earlier that analyses based on parameters of unwholesomeness provide data relevant to individual situations. This certainly applies to the batch of food under examination, though it needs a qualifying comment as far as the consumer is concerned. We can never completely assure the individual consumer that his meal will not make him ill, neither can the quality control microbiologist be certain that an individual purchaser will react unfavorably in the store. We are never certain of anything; at best we can only calculate the probability of an occurrence. Thus, data indicating that 80 percent of the people would show symptoms can be immediately interpreted by the consumer that he has an 80 percent probability of being ill if he goes ahead and eats. This alone is a much more meaningful deduction than he can make from a count of colony-forming units of an organism. Moreover, since the hazard is defined with greater specificity in this approach, it is not impossible to additionally warn persons liable to have subnormal tolerance for a particular parameter (as a result of impaired liver or kidney function, ill health or allergy, for example) that they are at greater risk. In principle, anticipating future encounters, an individual could even undertake to determine his own reactions to particular parameters (as dose/effect data for microbial toxins, for example). He would then have a very precise indication of his personal risk in any situation for which this type of microbiological data was available; however, this is really taking things too far.

Standards based on parameters of unwholesomeness lead to analytical results indicating the percentage of the population likely to show responses from the food and which are, therefore, rather unequivocal. In contrast, there will always be difficulties in relating numerical data to actual hazard, partly because more variables than just microbial numbers are important, and partly because the inherent indeterminacy of microbial counting methods leads to data of very arguable reliability. If we cling to numerical standards and analyses, this will continually plague us as a source of frustration and friction in the enforcement of microbiological standards. It is surely more convincing, irreproachable, ethical, or what have you for society (through the medium of its regulatory agencies) to enforce — and punish if necessary — the conduct of its food producers if it has the ability to unequivocally demonstrate potential hazards.

The directness of the parameter of unwholesomeness approach can be seen in other analytical applications. Thus, in principle, the approximate contribution of each step in a manufacturing process to the shelf life of the product can be calculated by measuring the incubation intervals needed to reach a specified unwholesomeness before and after that step (see the case of the comminuting step, described in Sequential Measurement of Amplitudes). In practice, only physical or mechanical manipulations of the product or those that do not introduce significant numbers of new microorganisms, such as deboning, comminuting, blanching, salt injection, etc., can be calculated so simply. The effect of steps where flora are combined, e.g. spicing, is the mathematical product of the individual effects from the mechanical mixing action employed (relatively constant) and the microbial quality of the additive (very variable).

Consider, for example, the path of one ingredient (say the pork in a skinless sausage line) in relation to the generation of a parameter of unwholesomeness (say, a complaint-provoking odor). Some of the processing steps experienced by the pork are

1. Live shipment received
2. Holding pens
3. Stunning/bleeding
4. Dehairing
5. Eviscerating
6. Blanching
7. Flaying
8. Hanging
9. Butchering
10. Deboning
11. Holding
12. Comminuting
13. Mixing (spices, seasonings, extenders)
14. Extruding (into skin)
15. Blanching
16. Deskinning
17. Weighing
18. Packing
19. Holding
20. Distributing

It would be feasible for the quality control microbiologist, using a sequential sampling apparatus, to determine the incubation intervals (under standard conditions) required for the partial pressure of an odor component over meat taken from each stage to reach a specified level. Steps (1) and (2) might, of course, be ignored.

Calculating the various ratios,

$$m_4 = \frac{I_4}{I_3}, \; m_5 = \frac{I_5}{I_4},$$

etc., would provide him with a set of values showing the *modulating* effect m of each processing step on product shelf life. For process steps having little effect on shelf life, m = 1. For deleterious steps such as comminuting, m < 1, and for beneficial ones such as blanching, m > 1. Steps where m is very much less than or greater than unity are the strongest shelf life modulators. Those where m << 1, in particular, may be prime targets in product quality improvement programs. If values for each step are available, those controlling shelf life or product quality might easily be picked out.

The quality control microbiologist will wish to minimize analysis times. Generally, these will be determined by the incubation intervals required. It will be apparent that in a process consisting of n steps, the probable shelf life of the product might be calculated from that at any intermediate step i, as follows:

$$I_n = I_i.m_{(i + 1)}.m_{(i + 2)} \cdots m_n \qquad (51)$$

If $m_i >> 1$ for the *i*th step in a manufacturing process, and the microbiologist has determined m for each of the remaining steps, he might conveniently make use of Equation (5:1) by carrying out routine quality control measurements just before this step rather than on the finished product. At this point, microbiological quality may be lower than at any other point in the life of a food until after it has been distributed. This procedure will provide the shortest analytical times. A trade must be made, of course, against reduced overall analytical precision since variances for each m add in the calculated value of I_n.

The picture would be very similar in other areas of factory microbiology. Again, the new method would be used in the same way as the old, but the implications of the data would differ. For example, the new data will say not that a surface holds 10,000 bacteria per square cm but that it reduces the keepability of the product by, say, 8 hours, or 15 percent.

The same directness can be brought to the investigation of sources in food poisoning outbreaks. Indeed, it often already is. Such investigations are complex procedures, of course, involving much more than just microbiological analyses. However, in situations where parameters of unwholesomeness have been defined — for example, in outbreaks of staphylococcal intoxication — direct analyses for the parameter are often made on remnants of suspect foods. Thus, demonstration of μg/g levels of enterotoxins in a food or characteristic physiological responses by test animals to extracts of it implicate a food more reliably than counts of *Staphylococcus aureus* organisms.

Such analyses fit the general framework of the parameter of unwholesomeness approach; they are, in fact, the first measurement (zero incubation time) of a SMART determination. The concentrations of toxins likely to cause obvious symptoms are, unfortunately, rather in doubt at the moment; the only values available have generally been estimated retrospectively from accidental outbreaks. Food poisoning investigations would benefit from increased motivation to produce reliable amplitude/response curves if the parameter of unwholesomeness approach became widely accepted.

When specimens do not contain detectable levels of toxins, analyses conventionally reduce to the making of microbial counts, followed by biochemical or serological characterization of suspect colonies. Implication of a food in the outbreak then rests on the probabilistic relation between parameters used to characterize the organisms, e.g. coagulase production, and those resulting in the hazard, e.g. toxin production, together with the probability of their multiplying in the specimen to yield an infective dose. Added to this is a good measure of experience (a priori probability). It should be obvious that sequential measurements of parameters of unwholesomeness in incubated specimens would often provide more direct evidence of their involvement.

Lightened Verification Burden

The need to ascertain the precision, accuracy, reliability, and limitations of techniques or instruments in several independent laboratories before they can be offered as bases for any recommended or official methods has been a severe obstacle to the development of commercial microbiological instruments (see The Verification Problem, Chapter 2). Small companies — the only ones currently interested in developing instruments for food microbiology — cannot sustain the financial drain of providing prototypes for such studies in anticipation of a market growth. Restructuring food microbiology in terms of parameters of unwholesomeness would shift the emphasis of the collaborative requirement and could attack its importance as an obstacle from several directions.

Much of the need for collaborative studies stems from the indeterminacy — and resulting irreproducibility — of conventional counting methods. It is extremely difficult to identify, let alone remove, sources of interlaboratory or intertechnician variation unless one can supply relatively permanent and reproducible analytical standards to each laboratory. The value of the collaborative study lies in the statistical suggestion that if the circulated instuctions for carrying out an analysis on a similarly circulated microbial reference specimen lead to similar data being returned from, say, six different laboratories, the method must be relatively tolerant of technician, reagent, and apparatus variations. The probability is, therefore, that the method will also yield comparable data in other laboratories. The ephemerality of microorganisms in foods as interlaboratory references, however, adds greatly to the problem of comparing or even accepting analytical results. For example, transporting a coliform contaminated ground beef standard from the coordinating laboratory to collaborating laboratories in different parts of the world, with all the unknowns of handling it may experience before the analyses are made, makes interpretation of any resulting data much less certain.

In contrast to living organisms, most parameters of unwholesomeness relate to intrinsically simpler and more predictable physical or chemical phenomena. They will be more easily prepared, duplicated, distributed, and maintained as reference standards than living organisms. In addition, since they pertain to physical, chemical, and biochemical measurements for which much instrumentation and technique has already been developed, factors affecting the reproducibility of analytical methods will, in general, already have been widely researched and cataloged. On both accounts, therefore, acceptance of the parameter of unwholesomeness approach may lead to increased reproducibility and reliability and to a reduced need for extensive collaborative study. A major part of the research effort for food microbiology will be directed at producing data on people, the onus of proof being shifted to government agencies, universities, and other institutions to demonstrate what is required rather than what can be achieved. In the limit, collaborative studies — as we know them at present — might become unnecessary. Thus, it might only be required of aspiring suppliers that they provide adequate evidence of their instruments' performances in indicating a standard parameter amplitude in the presence of specified sources of interference.

A move in this direction has, in fact, been taken recently by the AOAC which is encouraging association referees to describe methods in terms of *performance specifications rather than manufacturers' specifications*.[4] The burden to provide evidence of suitable performance of new automation in established AOAC methods will now lie with the manufacturer instead of with the association.

Specific developments for the parameter of unwholesomeness approach, such as the sequential sampling incubator, are relatively straightforward requirements for which performance should be relatively quickly evaluated. Elsewhere, its analytical methods will rely very much more on available instrument technology than our existing methods. Reduced development costs and the possibility of new items being rapidly accepted should both operate to the benefit of the small instrument manufacturer, who may more readily be able to offer apparatus on trial knowing that sale may follow fairly soon.

In addition, it will be noted that the larger instrument manufacturers have generally ignored food microbiology in favor of more obviously profitable areas of science. They would not, however, be long noticing the attractions of a market potential opened up by widespread adoption of this new approach. Serious interest by the larger manufacturers, who are much better equipped to support any financial burden incurred during collaborative studies or reeducation programs, could be of immense value in getting suitable instruments introduced into food microbiology.

Urgency Related Analysis Times

Universal desire for "instant" analytical data is very understandable but a rather tall order to meet in the area of food microbiology. Conventional counting methods require at least one day to execute because they depend on microbial multiplication. When successive subculturings to determine biochemical or other characteristics are involved, analysis times may run up to four or five days. One would hope a new approach might improve upon this; whether it can have the potential to return instant results for all possible microbiological analyses will depend very much on the kind of information it is asked to provide.

It is not impossible that increased research into methods of physically separating microorganisms from foods and into detection of substances that are relatively specific to them could eventually lead to analytical methods capable of detecting — within hours or minutes — many microbial species at levels pertinent to today's regulatory levels. Inevitably, the utility of such data would be impaired by uncertainties in the distribution of the analyzed material among organisms as they occur in foods. The real loss, however, would remain in the informative value of the data since, as has been emphasized (Chapter 4), microbial numbers alone do not adequately describe food quality.

Data that are useful mainly at the national level are surely not the most desirable and, unless there are other compelling reasons, we should not hold them as our analytical goals. Truly valuable data will primarily describe the quality of the specimens from which they were obtained. We are not, however, yet in sight of the day when the necessary information can be obtained by a deterministic approach. Any method predicting the combined number and multiplicative, metabolic, and interactive potentials of the organisms in a food will have to wait — in the foreseeable future — on the organisms themselves to demonstrate their abilities. In general, the prognosis must be that those who wish for the fastest data by deliberately avoiding microbial multiplication periods must also forego value. The point is emphasized here to anticipate complaints that the parameter of unwholesomeness approach also does not attempt to avoid an incubation interval so that it may provide little advantage in this respect over conventional counting methods.

In conventional numerative analyses, regardless of their concentration in the specimen, organisms are physically separated and growth has to begin again, as it were, from scratch. This procedure sets the detection level of organisms in all specimens down to that of the lowest. It does, at least, offer some advantage to the technician in leading to uniformly long analysis times for all specimens. However, quite frequently, organisms occurring in a food at a level that might almost have been detectable using a suitable apparatus are diluted down to one-millionth of their original concentration at the start of the analysis. The resulting loss of detectability is severe — it can only be restored with the aid of time and microbial multiplication. In respect to this, it should be noted that it is not uncommon for the environmental change experienced in being transferred from food to plating medium to retard the subsequent growth of microorganisms by several hours.

The parameter of unwholesomeness approach also leads to analytical methods based on microbial multiplication. In these, however, the organisms are neither mechanically disturbed, diluted, transferred to a strange medium, or otherwise interfered with, except, perhaps, in being favored with a more propitious temperature. It will be obvious that, in many cases, these organisms will have a head start in detectability over corresponding organisms in conventional analyses.

In SMART analyses, high quality specimens, i.e. those requiring microbial parameters to grow through many orders of magnitude before they reach physiologically active levels, will only show results in the fullness of time. In such cases, the SMART methods cannot be expected to provide significant time advantages over conventional analyses. However, incubation intervals and, therefore, analysis times *will* be short for specimens having great potential for generating parameters of unwholesomeness, that is, those requiring urgent attention. In the extreme — where a food is already unwholesome at the instant of sampling — the incubation interval will be zero, and the first measurement will indicate violation. In such cases, analysis times will be determined solely by measurement time and under suitable conditions could be classed as instantaneous.

In general then, the SMART methods, by providing analysis times related to food quality, will also provide results at a speed consistent with the urgency of the situation. This could be considered attractive in many areas of industry and health protection. Elsewhere, as in simple compliance work, analysis times will be those written into standards and may be shorter or longer than those in current use, as the case may be.

Equivalence as a Grading System

Analytical methods that are openly described as grading techniques tend to be eschewed by microbiologists as inferior to plate counts on the grounds of yielding less information about food quality. On the surface, for example, a simple and rapidly executed technique based on noting the number of decimal dilution stages passed through before colonies no longer appear on agar plates seems to provide much less precise a picture of microbial contamination than a conventional technique in which all colonies are enumerated on plates containing between 30 and 300 of them. Thus, the analytical answer — "less than 10^5/g" — appears to us much less satisfactory than the corresponding conventional answer — "3.7 × 10^4/g."

The subject can be treated from the point of view of information theory[3] where it is found that the information difference is debatable at best and minute at worst. Nevertheless, the superior image of counts persists. It would seem that to be acceptable, microbiological analyses should appear capable of providing answers either from a continuum or, at least, from a very large number of small incremental steps, in view of the fact that there are 100,000,001 possible answers when a food may contain between zero and 10^8 organisms/g.

It is possible to conceive of plate counts carried out using giant Petri dishes large enough to allow us to separate and count up to 10^8 colonies. Although such an analysis is impractical, its analytical precision would be very high. Alternatively, most probable number (MPN) determinations carried out using tens or hundreds of thousands of growth compartments are also capable, in principle, of high precision.[5] However, the conventional, practical plate count is quite incapable of deciding between all possible answers from zero to 10^8; it is, in fact, better classed simply as a fairly elaborate grading system. We cannot say, for example, that there are 7,394,218 organisms/g in a sample. We can only suggest that the true answer lies between 7.3 and 7.4 $\times$ 10^6. At the very best, if we assume counts are reliable to one decimal place, then they provide only 700 possible answers (grades) from samples containing between 1 and 10^8/g. If we more reasonably suggest data cannot be reckoned significantly different unless they differ by at least 20 percent, then plate counts provide only 101 possible answers or grades over the range from 1 to 10^8/g.

For simple compliance work, the number of grades an analytical system provides is irrelevant since only one answer (yes or no) is required.* However, when the quality or keepability of a sample must be determined, the number of possible answers the analysis is capable of distinguishing between, i.e. the quantity of information it provides, is important. With this in mind, we can look at the performance of analyses based on the parameter of unwholesomeness approach. Remember that we are talking about *quantity* of information here not *quality* (or value).

Compliance analyses can be dispensed with quickly; the parameter of unwholesomeness approach provides for exactly the same quantity of information as the count, i.e. a yes/no answer on compliance with a statutory standard. In practice, analytical precision will affect the reliability of data and reduce the actual quantity of information provided by a method. This effect will, of course, vary with the type of analysis and should generally be worse for microbial counts than for simpler parameters (see Indeterminacy, Chapter 4).

For quality measurements using the parameter of unwholesomeness approach, incubation intervals (the data) can range, in principle, from zero to infinity in steps as small as time is capable of being quantized. In principle, therefore, it can provide an infinitely greater number of grades or quantity of information than the plate count. In practice, of course, the quantity of

---

* In information theory, such an analysis provides 1 *bit* (binary digit) of information.

information will be limited by the maximum time over which samples can reasonably be incubated and the number of determinations it is economically feasible to make.

In a simple case, suppose the maximum residence time in the instrument is 48 hours and specimens are sampled every 5 minutes. The technique can distinguish between a maximum of $48 \times 60/5$ or 576 grades, a figure comparing favorably with the information content of a plate count analysis.

It is not actually necessary to make this many samplings during a SMART analysis in order to achieve the same efficiency. Indeed, even more precise data could be obtained using considerably fewer samplings since overall precision is regulated mainly by the frequency of the last few samplings. The actual means by which this would be accomplished is discussed in Chapter 6 (Programming Sequential Measurements). The important point here is that microbial analyses based on this approach appear to be capable of defining food quality at least as sharply as conventional counting methods. To change from one to the other would be, in this respect, simply to trade grading methods without losing — and possibly gaining — analytical exactitude.

Authenticity

When we wish to merely enumerate microorganisms in a food sample as accurately as possible, it is logical to transfer them from the sample to artificial growth media whose fertility or selectivity are more easily defined, and physically separate them to prevent mutual interaction between the growing colonies. Unfortunately, by so doing, we surrender the possibility of recording how those organisms would have fared in the habitat of the specimen itself. Of course, we can make educated guesses at some of the lost data. For example, we may already have determined that 95 percent of strains of the species looked for multiply slowly at the pH of the specimen. We may be less sure of the effect of the remaining flora on this multiplication rate, however. In the long run, our ability to supply part of the lost information statistically from experience allows us to use microbial counts for controlling food quality. It fails us mainly when we want to make judgements on individual specimens.

The parameter of unwholesomeness approach frowns on artificial dispersal of microorganisms and shuns artificial growth media in favor of the food itself. Provided a specimen is suitably representative, therefore, generation of any parameter within it is an authentic reflection of what will happen in the food. Effects due to variations in composition or integrity of the food and cooperative or competitive interactions between the flora are accommodated without need for guesswork. "What you see is what you get. . . ."

The Microbiophysiostatistical Research Incentive

The phrase may or may not be common, but its overall stridor must be familiar as physiological, sociological, and statistical reports proliferate. These reports are worktools for bureaucracies, and funding the studies on which they are based waxes attractive as we become increasingly concerned about our environment and human condition. All those concerned in any way with food quality or health might draw immediate profit from this gathering preoccupation by offering an eminently valuable subject for data acquisition — the construction of amplitude/response curves for microbiological parameters of unwholesomeness.

While such studies would obviously require considerable microbiological contribution in their organization and interpretation, they could be carried out away from the mainstream of food microbiology. The curves relate human responses to defined foodborne factors. Their determination, in calling for studies on people rather than microorganisms, belongs more to the realms of physiology, psychology, pharmacology, sociology, and environment than food microbiology itself. The data should be valuable not only for their immediate application to microbiological quality enforcement at local and governmental levels but also as fundamental measures of changing responses to education, socioeconomic or other advances, national, regional or ethnic differences, and the limits of consumer tolerances.

Constructing amplitude/response curves will require procedures somewhat removed from the traditional methods and attractions of food microbiology. The labor and expense involved need not intimidate us if we feel this is the way to go, for the major burden need not rest on microbiologists. The necessary forces are well prepared within other sciences and if funds became available could, I am sure, mobilize with enthusiasm for a cause of such demonstrable value as the potential amelioration of national health.

Data Format and Emotive Impact

The parameter of unwholesomeness approach certainly encompasses a number of long-developed detection procedures that have so far failed to become popular in microbiological analysis. Many of these, perhaps, because of practical or interpretational difficulties did not deserve to become popular. However, it is not always the case. I believe the poor psychological impact of their data, relative to numerical germ data, has had much to do with the failure of some methods to achieve a rightful acceptance.

The human aspects of scientific methods are so important to their acceptance — particularly in microbiology — and so generally ignored in the scientific literature that I have devoted Chapter 7 to a more detailed discussion of the subject. However, the point is raised here in order to introduce what may be another significant attraction of the parameter of unwholesomeness approach, namely, that the images its data may be persuaded to kindle might compete effectively at the subconscious level with those projected by numerical standards and methods. Some possible forms and terms will now be brought in. The following section is just for play but is meant to have a serious undertone.

Projection 2,000

11:30 AM   The scene is the regional **BIOSCREEN** laboratory in a national regulatory agency. In one corner, a microbiologist discusses hygiene with a restaurant manager over the telephone. At the side of her workbench is a console where a video monitor, disregarded for the moment, periodically flashes analytical data about samples under analysis. The laboratory is quiet, and there is little movement.

<u>11:31 AM</u>   A courier arrives with ten packages of sliced ham, picked up during routine surveillance. The biologist breaks off her conversation and signs for the samples. As she carries them to her workbench, she glances at an illuminated display on the wall. It reads 13. With thirteen specimen locations available in the laboratory's analytical system, all the new specimens can go straight in.

<u>11:33 AM</u>   The biologist flames a scalpel and empties ham from the first container into a specimen vial. Turning to the console, she presses a key marked SPECIMEN. In front of her the monitor writes

INDICATE WHICH BANESPECS YOU REQUIRE
A. EH
B. H2S
C. NH3
D. ACID
E. E. COLI
F. C. PERFRINGENS
G. STAPHYLOCOCCAL TOXINS
H. SALMONELLA

She indicates using key G, and the monitor writes

PLACE SPECIMEN IN LOCATION 17.

The microbiologist obeys and types DONE. If she had indicated any of the analyses A through D, the instrument would first have asked her to insert calibration standards in Location 17. Now the screen says

RECORD SPECIMEN DATA NOW. TIME AND DATE ARE ALREADY RECORDED.

As she types, the monitor displays the name and address of the source.

<u>11:35 AM</u>   The screen says

THANK YOU. ANALYSIS IS PROCEEDING.

During the next few minutes, the microbiologist inserts the remaining specimens into other locations in the instrument in the same way. She carries old vials to the preparation room for disposal; at the moment it is deserted, but the part-time steward will be in again tomorrow afternoon.

11:50 AM With ten analyses for ability to produce staphylococcal enterotoxins under way, the microbiologist returns to a manuscript she has been preparing. The instrument will incubate the specimens and make radioimmunoassays of toxin concentrations during the next 12 hours.

Long ago, the regulatory agency determined levels of toxins at which 0.1 percent of humans experienced some effects. A standard was then laid down to the effect that ham must not generate this physiological threshold concentration in less than 12 hours incubation at 40°C. If a specimen meets the standard, the instrument will silently print out

SPECIMEN . . . OK

12 hours from now, if any specimen is more active than the standard allows, the instrument will let her know as soon as it finds out.

1:30 PM Immediately after lunch, the microbiologist notes that the instrument has "OK'd" several weiner specimens placed in it the day before from the point of view of their Eh, H_2S and NH_3 production. These tests replaced the old plate count; while not regarded as being of particular significance as hazard indicators, they are frequently carried out following consumer complaints.

2:45 PM An alarm on the instrument begins to whoop. The microbiologist puts down her manuscript and notes that the monitor is flashing

SPECIMEN 3. STAPH. TOXINS ALERT. ATTAINABLE LEVEL EXCEEDS REGULATORY THRESHOLD. INDICATE SPECIMEN REMOVED FOR FULL DATA PRINT-OUT.

Location 3 contains one of the morning's ham specimens. She removes the vial and types in DONE. At once the alarm falls silent and the instrument prints out BANESPECS — complete data taken at decreasing intervals, showing rapidly rising toxin concentrations in the specimen. The microbiologist begins to act on the ALERT.

The BIOSCREEN system provided information showing that regardless of the organisms or their numbers in the specimen, its overall activity was sufficient to yield toxins capable of causing staphylotoxicosis in 0.1 percent of humans after only 3¼ hours at 40°C. This violated the regulatory standard of 12 hours and caused the instrument to signal that the staphylococcal toxin Attainable Level Exceeded Regulatory Threshold (staphylococcal toxin ALERT). The data printout was used in confronting the ham supplier with unequivocal evidence that his product was unacceptably close to causing illness.

That was science fiction. It was also a logical projection of the approach to microbiological analysis described in this book. Within a few years, food microbiologists could have interactive instruments like the one described. Early instruments would probably make determinations relevant to organoleptic responses (spoilage). Developments for analysis of toxins and infective agents could follow fairly quickly as the necessary data bases and analytical procedures became available.

One obstacle stands in the way. The concept on which these potential microbiological instruments are based entertains little respect for microorganisms themselves in foods. Could such instruments and analytical procedures be widely accepted? I do not know. I am sure of only one thing — we are unlikely to see food microbiology evolve into a precise, properly instrumented science as long as we continue to describe the quality of our food by the numbers of the various organisms we can find in it.

REFERENCES

1. Stahl, W. H. (Ed.): *Compilation of odor and taste threshold values data.* ASTM Data Series DS-48. Philadelphia, American Society for Testing and Materials, 1973.
2. Venn, S. Z., Woodburn, M. and Morita, T.: *Staphylococcus aureus* S-6: Growth and enterotoxin production in papain-treated beef and ham and beef gravy. *Home Economics Res J, 1:*162, 1973.
3. Sharpe, A. N.: Some theoretical aspects of microbiological analysis pertinent to mechanization. In *Mechanizing Microbiology*, (Eds.) A. N. Sharpe and D. S. Clark. Springfield, Thomas, 1978, p. 19.

4. Association of Official Analytical Chemists: AOAC Board of Directors approves new automated methods policy on performance specifications. *The Referee, 2:*1, 1979.

5. Sharpe, A. N. and Michaud, G. L.: Enumeration of bacteria using hydrophobic grid-membrane filters. In *Mechanizing Microbiology,* (Eds.) A. N. Sharpe and D. S. Clark. Springfield, Thomas, 1978, p. 140.

Chapter 6

SOME PRACTICAL CONSIDERATIONS

NEW IDEAS may sound simple, but putting them into practice is rarely straightforward. I'm sure the parameter of unwholesomeness approach would be no exception if it were adopted. This chapter discusses a mixture of the most obvious practical problems, background details bearing on its use, and some possible formats for microbiological instruments of the future. Doubtless, different readers will find other aspects equally important or obvious.

One should keep in mind how the approach leads to two rather dissimilar types of microbiological analyses. On the one hand, compliance analyses pertinent to quality enforcement may be very simple affairs, consisting of single comparisons against a standard at the end of specified incubation intervals. Instrumentable by preference, they could also be easily carried out manually if it were necessary. On the other hand, keepability analyses, for monitoring or predicting shelf life, call for sequential measurement (SMART) techniques that are not likely to be considered suitable for manual execution. At present, there is little difference between these two types of analysis, save perhaps, that quality control laboratories may ignore laborious details of official or recommended methods in the interests of throughput. Small laboratories may monitor the overall running of a plant but because of the labor burden involved may be unable to reassure themselves adequately about the product's compliance with statutory standards. In the new approach, this is the simplest measurement; quality control laboratories may thus compete with the regulatory agencies in analytical rigor, even if they can do nothing else.

Except for very fundamental problems, such as how we can define the various parameters or measure amplitude/response curves, this chapter will be found to dwell mainly on aspects of SMART techniques. The bias results from this very asymmetry.

Once the principles have been laid down and the form of compliance analyses decided, there is little else to say about them. Specimens either comply or they do not, under the standard incubation conditions. Scope for ingenuity and discussion lies only with the SMART analysis.

REPRESENTING FOOD QUALITY BY PARAMETERS OF UNWHOLESOMENESS

Not the smallest problem in adopting this approach to food microbiology would be that of defining many of the parameters on which its techniques must rest. It is, of course, particularly important to define hazardous parameters adequately. However, the general care with which we assign parameters will bear strongly on the relevance and utility of subsequent microbiological analyses and the ability of the approach to withstand criticism. We can expect wide variations in the ease and confidence with which we make these assignments.

When we attempt to describe the microbiological quality of a food, we are effectively placing limits on those anticipated problems for which we have analytical data. I doubt we shall ever have the capacity to obtain *complete* microbiological information about a food, i.e. be able to predict every possible physical, chemical, biochemical, or immunological state to which it might progress. Somewhere, we make trades between desirability and feasibility and, as a result, between analytical specificity or generality. There are notable differences between our overall appreciation of a food by the parameter of unwholesomeness approach and many conventional microbiological criteria.

The first stages of routine counting techniques are like rather rounded tools, yielding an indefinite sort of contact with the workface over a comparatively dispersed set of specimen prognoses. For example, while the test for coagulase-positive staphylococci does suggest the possibility of encountering enterotoxins, it cannot, because of the very wide range of enterotoxigenic activities that exists, provide assurances of imminent staphylococcal poisoning. More finely honed analyses, from which the activity might be inferred, must usually be left until later or ignored.

In contrast, many analyses for parameters of unwholesomeness can immediately be extremely specific. Used without care, they might sometimes be too specific, providing us with very precise answers for too narrow a range of conditions and leaving us in the dark about closely related questions. For example, analyzing specimens for their ability to generate physiologically active levels of staphylococcal enterotoxin C_2 would provide us with very precise information on the dangers of consumers becoming ill from that toxin but very little about other, equally probable, types of staphyloenterotoxicosis.

Since it will hardly be feasible to follow the generation of every related toxin, the practical requirement must obviously be for a much less specific analysis directed at a more general feature of the toxins. Cross-reacting antibodies do exist or can probably be prepared for many of these situations. It is already possible that just two cross-reacting antibodies, for example, may suffice for the determination of all of the known (and possibly many unknown) staphylococcal enterotoxins. Without this sort of generality, parameter of unwholesomeness data must suffer the same criticisms of information deficiency levelled against the microbial counts. Very specific analyses are, however, valuable in tracing the path of particular strains, for example, during the investigation of a food poisoning outbreak.

Difficulties of defining parameters relevant to keepability or spoilage result partly from the fact that the visual, olfactory, tactile, and other cues causing us to pass over food or otherwise reject it as unwholesome or unsavory are very complex and capable of interfering with one another synergistically or antagonistically. On different occasions, for example, rejection of ground beef may be primarily caused by stimuli from its appearance, odor, taste, or feel; on the other hand, our response may be due to various combinations of these, in a manner that cannot easily be related to their individual thresholds of perception.

The course of deterioration of a food is rarely completely predictable and, in a given situation, any one of several parameters might be capable of changing to become the predominant stimulus. Thus, even though our reaction may be stimulated by the intrusion of a single parameter into our sensibility, it might equally have been triggered by others had they not happened to have developed less rapidly in that particular specimen. Unless we are overwhelmingly more likely to be stimulated by one parameter than by others, therefore, keepability analyses should generally seek to follow the development of several relevant parameters. Those we actually define as being worth measuring will depend, of course, very much on the food and our interests as manufacturer or guardian agency.

There are two stages to approaching keepability determinations. First, we need to decide what general features of the food are most relevant to our perceptions. For example, our perceptual cues for avoiding milk may result from the expression of one or more different phenomena, such as its viscosity, ropiness, acidity, bitterness, clotting, and odor. These can be redescribed in more scientific terms as bases for shelf life analyses.

Similarly, expressions of deterioration such as color change, weeping, sliminess, structural integrity, and off-flavors stimulate us to avoid stored meats. These may also be described more scientifically as bases for analyses. If they are all important, we should try to measure them; the first four, at least, are relatively simple requirements, and the last — if we are reasonably selective — need not present an insurmountable problem.

If this sounds formidable, it should be remembered that plate counts and similar analyses alone are not good predictors of food deterioration because each of these changes may be accelerated differently according to the type of microbial contamination chancing to be present. In fact, basing any assessment on measurement of a single parameter has probably been a major cause frustrating earlier attempts to introduce alternative approaches to quality determination. For example, complaints such as

> While over 40 different methods have been proposed for detecting incipient spoilage . . . the lack of general agreement on the validity of any of these methods reflects in large part our lack of agreement on a precise definition of microbial spoilage. . . . The most urgent problem that confronts us in this whole area is the one of identifying what it is that we seek to avoid in refusing to accept spoiled meats as desirable products.[1]

seem to me to be mainly comments on a general reluctance to

introduce or accept multiparameter measurements. Humans are complex, multifunctional detectors, and there is no uniquely definable thing we seek to avoid, in meat or any other food. Moreover, we have no basis whatsoever for expecting the different possible changes to correlate with one another or with any other single phenomenon because the driving mechanism behind them has almost infinite variability.

There is little point, for example, basing an assessment of quality on just hydrogen sulfide generation if there is any possibility of specimens not containing significant quantities of organisms able to produce this chemical. We would run the risk of specimens passing the test while the food itself reduces to a festering jelly long before it reaches our predicted life. The same applies to measurements of water-binding capacity, pH, dye reduction, and any other single parameters proposed over the years as bases for assessing quality. Anything less than the broad perspective is not enough.

A second step may be involved once the relevant parameters — if they are odor or taste stimuli — have been roughly defined. Almost invariably, a number of related compounds is responsible for our olfactory or gustatory perception of deterioration. Our perception of odor, in particular, is rather specific both to each compound and to the mixture as an entity. When no single compound appears to play a predominant role, we may (as in the staphylococcal toxins example) need to trade discrimination for generality at some loss of precision. For example, we might measure total volatile amines, acids, alcohols, esters, etc., rather than individual contributors to an odor. Combining contributions this way from a group of related compounds will blur the implications of the agreed threshold and reduce the value of the analysis. Much research has been directed towards objectively simulating the reactions of the average human nose and, in principle at least, this feat may be possible using, say, suitable analyzer/computer combinations. However, since we are only concerned here with organoleptic acceptability and not human lives, the cost of developing the necessary instrumentation solely for SMART analyses would probably be out of proportion to the value of the result, and we should not assume that it will become available quickly from other sources.

METHODS FOR GENERATING
AMPLITUDE/RESPONSE CURVES

The essence of this future approach to food microbiological analysis derives as equally (or more so) from people as from microorganisms. Amplitude/response curves are interfaces between the two — human expressions of the only truly significant consequences of microbial contamination. It is, therefore, vital for methods by which these curves are generated to be beyond reasonable criticism if the approach is adopted.

There is nothing new in the measurement methods alluded to here. Many of the techniques employed are as old as or older than food microbiology itself and, for this reason, are not described for their practical details. Any deviation from conventionality rests entirely with the emphasis placed on using such data to define microbiological, analytical objectives and the instruments or methods to which this viewpoint gives rise. This section covers what seems to me to be the more valuable aspects for further discussion.

The subject calls for contributions from many areas of science. A fundamental need for microbiological expertise in describing food quality characteristics will not change under the parameter of unwholesomeness approach. However, there will be a need to de-emphasize microorganisms as the natural center of attention, regarding them as of interest simply as prime movers in the development of more immediately relevant parameters, in many ways subordinate to, in fact, but forming the link between toxicological, physiological, environmental, and other interests. While producing data to generate curves is quite another activity to setting threshold values, it cannot be completely separated from it so that commercial, political, and legal interests must have some influence if only on the meticulousness of the resulting research. Production of amplitude/response curves will thus be very much an interdisciplinary affair.

Methods for obtaining suitable data fall into three categories. At the lowest level of experimental difficulty are the organoleptic parameters relevant to spoilage. Being quickly perceived and of low or zero hazard, they present the least problems of obtaining volunteers for assessment panels and in identifying and recording results. Sensory evaluation studies of exactly the type required are made routinely in research laboratories of many large food manufacturing companies, consulting organizations, and other institutions.

From this to the next level — the generally nonfatal food poisonings or foodborne infections conventionally associated with contamination by and/or proliferation of *S. aureus, C. perfringens, Bacillus cereus, Shigella,* and *Salmonella* species, etc. — involves a considerable increase in experimental difficulty. Among problems long encountered in this area are those of obtaining sufficient and suitably representative volunteers, or observing delayed reactions, and of satisfactorily relating apparent reactions to the stimulus in the presence of other interfering stimuli and reactions (analytical noise).

The most difficultly accessible data unfortunately relate, of course, to the very hazardous toxins or foodborne infections. The legal, moral, and logistical problems of obtaining satisfactory human dose/response data are certainly great and may be reinforced in some cases by problems resulting from the very low analytical detection limits required. There is no doubt that this would be the last area of food microbiological analysis to be considered for the parameter of unwholesomeness approach. For the time being, the best that can be hoped for is probably that increasing enthusiasm for the approach in other areas might stimulate interest in funding and discussing the problem.

Worms, protozoa, and other parasite contaminations should, in principle, be included in the approach. However, while their detection by, say, immunological methods or direct observation is made successfully in specialized laboratories, instrumental considerations seem so low in priority, at least in the more developed countries, that I do not propose to worry about them. In any case, contaminations such as these, along with viruses, being incapable of self-amplification within foods, occupy the limiting situation where the parameter of unwholesomeness approach becomes indistinguishable from the conventional.

Spoilage Parameters

The diversity of aspects and objectives in sensory testing makes general commenting difficult. From the point of view of consumer complaints, loss of sales, etc., keepability is very much of a commercial rather than a regulatory interest. Any burden of research is thus likely to devolve on food manufacturing companies and/or their consulting institutions. Acquisition of data on which to base SMART analyses may be directed at national responses in the case of some widely retailed foods or very much at a regional or ethnic level in others. The scale, composition of sensory panels, and general thoroughness with which a survey is carried out will depend on these factors, on the size of the sponsoring organization, and the financial implications of the resulting data base.

The American Society for Testing and Materials publishes excellent introductory books on the principles and practices of sensory evaluation.[2, 3, 4] Most of the normal rules for providing suitable physical conditions, selecting and training panels, maintaining the physiological sensitivity of panel members, preparing or presenting samples, and, in particular, the psychological control of tests and panelists apply to the type of panels required for the determination of amplitude/response curves in SMART data bases. The required analytical method will presumably be close to what are known respectively as the Hedonic Scale Method[4] and the Food Action Scale Method.[5] In the former, samples are presented in succession to the subject, who rates them according to categories such as

Like extremely
Like very much
Like moderately
. . .
to
. . .
Dislike extremely

In the latter, samples are also presented successively but are rated into categories such as

I would eat this at every opportunity.
I would eat this very often.
I would frequently eat this.
. . .
to
. . .
I would eat this only if I were forced to.

For parameter of unwholesomeness work, acceptance categories might be as simple as

I would buy/use/eat the product.
I would not buy/use/eat the product.

or be additionally graded, for example, by

I would buy it at $2.99 per pack.
I would buy it at $1.99 per pack.
. . .
I would complain to the store manager.
. . .
I would avoid all this company's products in future.

if data are relevant to point-of-sale decisions by the consumer, or

I would eat it raw.
I would fry it first.
. . .
I would take it to the Public Health Laboratory.

where data relate to domestic responses.

To stimulate panelist's responses, parameters of unwholesomeness do not need to develop through microbial activity. Amplitudes can be varied artificially from sample to sample in an otherwise uniform and generally acceptable product base. For example, tactile and rheometric properties of ground beef may be modified by plasticizers or proteases, its color by dyes, its odor by sulfides, fatty acids, etc., at suitable pH, or its taste by suitable peptides and other agents.

Wherever possible, realism should be maximized by running tests inside a grocery store or asking panelists to consume the food. Presenting successive samples, paired comparisons, triangle tests, and other techniques designed to facilitate statistical treatments will usually reduce realism and should be avoided.

Nonfatal Toxic or Infective Parameters

Procedures to satisfactorily determine human dose/response relations for the (generally) nonfatal parameters in food are exactly those used in general and industrial toxicological investigations or during preclinical and clinical evaluation of drugs. The logistics, methods and clinical problems are well established.

Legal, moral, and clinical problems of evaluation would seem to be less severe for these natural toxins and infections since symptoms, side effects, and chances of recovery are already well documented. That suitable data do not already exist presumably reflects a lack of adequate funding and, ultimately, the fact that these materials are not deliberate food additives or sources of immediate profit like pharmaceuticals and have not, therefore, aroused the same human sensitivities or emotions.

The slope of a dose/response curve depends on the standard deviation of response data, and standard errors are minimal at the 50 percent response level. When animals are used (predictive toxicometrics), effects of injurious substances are frequently reported as ranges or as ED_{50} (activity), TD_{50} (toxicity), or LD_{50} (lethality) values, these latter indicating the doses at which 50 percent of test animals showed the appropriate resonse. The corresponding microbiological value is ID_{50} (infectivity), quoted as the number of colony-forming units resulting in this percentage of clinical infections.

For humans, much lower doses (say, ED_1 or smaller) must be determined; the various values are sometimes described as maximum allowable concentration (MAC), threshold limit value (TLV), or minimum infectious dose (MID). In animal experiments, values for ED_1 may be readily estimated for equations used to determine ED_{50}; the ED_1 has a high standard error and little statistical significance unless large numbers of subjects are employed. However, it becomes extremely important as a guide for maximal safe doses for clinical trials in man.

In the long run, reliable human dose/response data must come from humans. If we want suitably precise data, there seems to be little alternative (statistically) to using many subjects. There are, at present, very large gaps in our knowledge. A little data, varying widely in detail and significance, is already available regarding physiologically effective doses of some toxins and infective microbes. Most has been accumulated through investigations of natural incidents. A smaller quantity, though probably of much greater relative value, has been obtained through experiments with human volunteers.

Thus, clinical illnesses have been reported following ingestion of 1 to 25μg of staphylococcal enterotoxins,[5] of 10^9 to 10^{10} cells of *Shigella flexnerii*, of 10^8 to 9×10^9 cells of *Escherichia coli*, of 2.2 $\times 10^{10}$ cells of *Streptococcus faecalis*, of 3 to 5×10^{10} cells of a *Proteus* species; of 10^5 cells of *Salmonella newport*, of 5×10^6 cells of *Sal. anatum*, of 5×10^7 cells of *Sal. meleagridis*, of 1.3×10^9 cells of *Sal. pullorum;*[6] and of 4 to 6×10^9 cells of *Clostridium perfringens.*[7] Owing to the great variation in human responses to any particular dose, the small number of subjects in all but the *Salmonella* investigations makes the value of even these data doubtful.

The variability results from factors such as previous exposure, age, nutritional status, general physical and mental condition, acidity of the gastric juices, and probably the vehicle in which the organisms occur. It has recently been suggested that this last may be rather important, in the light of evidence of clinical infections occurring after eating chocolate that provided doses of only 100 cells of *Salmonella eastbourne.*[8] This dose, it will be noted, is some 10^4 to 10^6 times smaller than doses reported for other strains of the organism in different vehicles. Similarly, it has been suggested that when ingestion occurs a few hours after a meal, far fewer cells may provide a clinically infective dose.[9]

If these suggestions are correct, infectious microbes may prove rather difficult to handle by the parameter of unwholesomeness approach for some time. Variability itself is not a problem since the methods are directed at measuring percentages of responses in a population.* Rather, situations where a very high infection rate results from a dramatically lower dose at a critical time of ingestion may pose severe problems of interpretation and concurrence between parties responsible for drawing up standards. Considerable emphasis is likely to be placed on this time of least resistance. If the proportion of a population in this high risk condition at any instant is higher than the percentage agreed to be acceptable in the standard, this will completely dominate the value arrived at for the infective dose. Analytical limits of detection subsequently required may be so low as to be beyond the capability of present immunological or serological procedures; for many years they might be realized only by conventional plate counts and similar methodologies.

---

* Although, the standard deviation in a population affects the number of subjects needed to determine a response curve to any given accuracy.

On the other hand, the apparent ability of very small doses to cause infection does not necessarily prove the existence of occasional special protective effects as organisms pass through the stomach. Doses reported for most organisms are those causing a very significant proportion of responses in subjects (usually convict volunteers). They relate, therefore, to the middle or upper portions of dose/response curves. In the case of the illnesses resulting from chocolate, the percentage of responses was unknown; certainly, from the number of reported cases and the probable number of consumers, it was very small. The illnesses that did occur may, therefore, be simply regarded as a normal expression of the extreme end of a quite normal dose/response curve. My feeling is that there just is not enough data available from which to form a useful opinion. Whether one considers infectivity from the conventional or the parameter of unwholesomeness viewpoint, questions of the effect of the vehicle and the time of ingestion on infection rate can only be satisfactorily answered by obtaining more detailed dose/response data.

Clinical or toxicological investigations on human volunteers may commence with very low doses of organisms or toxins, gradually increasing until the percentage of detectable responses over the normal level of physiological fluctuations reaches or approaches that agreed upon in the standard. Presumably, a response would generally be taken to be any overt symptoms characteristic of those caused by the parameter of interest, regardless of its severity. The appearance of similar symptoms at the zero dose level (analytical noise) will considerably affect the size and duration of any experiment. In this respect, prior determination of dose/effect data (the relation between the dose and magnitude of the effect in individuals) might be valuable. Biomedical studies — for example, examination of disturbances in levels of enzymes or other blood constituents — may allow dose/response data to be obtained at levels lower than those giving rise to overt symptoms and may also minimize the analytical noise level.

Very Hazardous Parameters

"In fact," said Mustapha Mond, "You're claiming the right to be unhappy."

"All right then," said the Savage defiantly, "I'm claiming the right to be unhappy."

"Not to mention the right to grow old and ugly and impotent; the right to have syphilis and cancer; the right to have too little to eat; the right to be lousy; the right to live in constant apprehension of what may happen tomorrow; the right to catch typhoid; the right to be tortured by unspeakable pains of every kind."

There was a long silence.

"I claim them all," said the Savage at last.

Mustapha Mond shrugged his shoulders. "You're welcome," he said.

Aldous Huxley, *Brave New World.*

Without doubt, this must be the most contentious application area for the parameter of unwholesomeness approach. I write on it with great timidity, for the major issues are far beyond microbiology. What is, in principle, possible from the microbiological, toxicological, and clinical points of view, is almost irrelevant. Other aspects — the moral, ethical, legal, religious, sociological, and political considerations alone, in a society where human rights (whatever that means) are to be respected — frankly daunt me. How *do* we obtain satisfactory human data?

Actually, of course, whether we stay with the numerical approach or not, we need better human response data. Whatever our approach to food microbiology, we ought to face these same considerations. If we do not, we are avoiding the true problems. It seems to me that in view of its importance, remarkably little work on the quantitative relation between microbes and illness is currently under way. Could it be that the intense efforts we see to detect, isolate, and characterize microorganisms in food microbiology are actually diversionary acts in the face of the great problems of working with people?

The main hope for the future seems to be that research in predictive toxicometrics will continually improve the pertinence of animal studies. This is currently the only avenue open to microbial toxicological studies for the very hazardous parameters such as the botulinal and aflatoxins. For these, at least, useful predictive data may be expected sooner or later. Aflatoxin and other fungal toxin concentrations are, in fact, already used as quality descriptors for foods. Unfortunately, when we move from toxins to infectious organisms, animal studies are far less valuable.

Ultimately, however, even the best predictive data require validation if they are to carry any authority whatsoever. One can argue that very minimal effects and very low response rates will be tolerated for the very hazardous parameters. With the availability of antitoxins, antibiotics, etc., to rapidly arrest toxicoses or infections, the risks involved in properly conducted biomedical and clinical studies on humans could, therefore, be very small. At least, that is, where there is no suspicion that longer term effects like carcinogenesis are associated with a parameter.

There are, of course, increasing obstacles to experimentation with human subjects, even if volunteers will come forward and even if such persons can, therefore, be assumed to be representative of the remaining population. The traditional overt transaction with imprisoned criminal offenders seems to be frowned on these days. Undoubtedly, clandestine pharmacological (and probably biological) studies are carried out on unwitting subjects, as we have witnessed with the CIA. But the data from such experiments, as with those on prisoners of war, are unlikely to be useful to food microbiologists, even if they were to be publicly disclosed. We would not, of course, want any suspicion of such stealth to be associated with food microbiology. It is essential that we be seen to respect the rights of individuals even if, in consequence, many unknown individuals are condemned with "the right to be tortured by unspeakable pains of every kind" for longer than is necessary because of our lack of knowledge of microbial interactions.

It is a very difficult area, to be sure. Serious discussions of the parameter of unwholesomeness approach would, as it were, bring its controversies to a head. But I cannot accept that the problems raised in it are unique to the parameter of unwholesomeness approach.

SETTING TOLERANCE LEVELS

An adequate discussion of scientific, moral, ethical, social, legal, economic, and other factors bearing on the determination of tolerance amplitudes for given parameters and populations itself requires a larger book than this. The arguments and emotions raised for the problem of rationally balancing risks and benefits, which are currently very evident in the area of food additives like sweeteners and colors, apply in their entirety to microbial sources of hazard or complaint.

Permitted levels must be a compromise between opposed viewpoints. At one extreme, consumers — or rather, "representatives" exploiting consumers' fears — may press for absolute safety (zero tolerance). At the other, evanescent manufacturers find microbiological quality considerations quite unprofitable. In between, assessing tolerance levels for microbiological agents has traditionally been a matter of rational compromise between regulatory agencies and the more responsible manufacturers. Finite levels of all hazards or complaints must be accepted, if only because of the impossibility of proving absolute safety. Zero tolerance — insistence on totally negative analytical results — is recognized scientifically as an impractical concept, continually confounded by improvements in analytical limits of detection. Most food manufacturers, however, even if only through the necessity of staying in business, also recognize the undesirability of leaving unnecessarily high potentials for complaint in their products.

A toxicological procedure used extensively in the past for establishing acceptable daily intakes or maximum allowable concentrations of additives has been to apply a safety factor (commonly 100) to the experimental level causing no observable adverse effects in suitable animals. More recently, there have been moves towards determining levels in exactly the manner required for microbiological purposes; that is, by assessing risks through mathematical extrapolation of dose/response data and allowing "informed" consumers or their representatives to decide acceptable risk levels.

A great deal of retrospective microbiological data on, for example, the levels of organisms or toxins encountered in various foods across a country and on what can or cannot be achieved by reasonable manufacturing practices has already been collected as the basis for the standards, specifications, and guidelines we now have. The quality of such data varies greatly, in particular, through variations in reporting illnesses and the difficulties of determining toxins or microbial numbers after the event. Nevertheless, sufficient information is available, with sufficiently established mechanisms for gathering more (in the form of monitoring laboratories, food safety committees, councils, etc.) to serve as guides for the initial formulation of quality objectives.

Determination of acceptable risk levels should not be in the hands of one sector but should be a societal decision achieved through delegation of representatives from all reasonable areas of interest. There is also need for a relatively permanent body to oversee the activities of individual panels so that microbiological quality measures may be treated in an integrated manner. It may be that this cannot be achieved within the considerations of existing foods and drugs acts. The situation for spoilage parameters is simpler and more flexible; the objectives and the depths of interests consulted in setting them will be very much more an executive decision.

As with food additives, satisfactory quantitation of risk/benefit for microbiological parameters is very difficult. Even if risks themselves can be assessed using human volunteers, the necessity of or the benefit conferred by consuming a given food is rather unquantifiable. The consumer, for example, without significantly reducing the nutritive quality of his diet, can reduce his risk from one or more foods to zero simply by avoiding them altogether. However, if he avoids too many foods, he becomes significantly at risk from other ills. Such problems emphasize how treatment of the subject should be integrated rather than being attacked piecemeal. Typical data required by delegated panels will be

1. the nature of the risk
2. the nature of the population at risk
3. assessment of its present exposure in relation to current manufacturing, handling, etc. practices
4. the population's apparent requirement for the food, for example, as a monetary choice, relative quantity consumed, frequency of consumption, etc.

With a reasonable abuse condition and maximum acceptable response level agreed upon, definitive toxicological, clinical, or organoleptic evaluations of amplitude/response can begin, using this tolerance level as the upper response objective. During the subsequent interval in which amplitude/response data are being generated, any reassessment of both the tolerance limit and defined abuse condition can be made. Standard statistical protocols will allow uniformity in deriving tolerance limits; agreement on these will presumably be a responsibility of the permanent overseeing body.

Assuming that laws have been altered to provide the appropriate translation mechanisms, such societal decisions can presumably progress to eventually become statutory standards. Whether a chaotic adjustment period accommodating dual standards (such as has been evident during conversion to the metric system in most countries) can be envisaged for situations where numerical standards already exist is open to question. There also remains to provide for reviewing and updating such standards in the light of, say, future retroactive data or fresh evaluations of dose/response relations with changing habits of food preparation, handling and consumption, the overall health of the population, the economy, and so on.

Finally, there may sometimes be the matter of deciding which effects of a parameter should be considered in the setting of tolerance levels. For example, at the present, tolerance levels for staphylococcal enterotoxins would almost certainly be defined in terms of their emetic effects if we were to lay down standards for them. However, the possibility exists that these toxins exert immunosuppressive effects at doses thousands or millions of times smaller than those for emesis. A decision, at some future instant, to control toxin levels in foods on the basis of immunosuppression rather than emesis would make their detection at physiologically active levels much more difficult. It should be noted, however, that in the foreseeable future, the problems of enforcing food manufacturing practice to achieve this level of quality — regardless of whether it is viewed as counts or as parameter amplitudes — would probably prevent the use of such stringent standards.

PROGRAMMING SEQUENTIAL MEASUREMENTS

When all other sources of sampling, reagent, and instrumental fluctuations have been accounted for, the precision of a SMART datum is determined by the interval between readings as the parameter amplitude passes through its defined physiological threshold. The more rapid the sequence of analyses, the greater the precision. In Chapter 5 (Equivalence as a Grading System), I described how a SMART analysis period of 5 minutes could provide a grading capability comparing favorably with a conventional microbial count.

If we had to handle many samples, and if reagents were costly, this could lead us to somewhat cumbersome and expensive analytical systems. We do not, however, need to sample constantly during a SMART analysis in order to achieve its inherently high precision. The data can be obtained from considerably fewer samplings than the final rate implies since we can easily assure ourselves that a particular parameter will not increase faster than a certain rate. Bearing in mind that the parameter of unwholesomeness approach will facilitate instrumentation and that SMART instruments will almost certainly be controlled by computers or microprocessors, it is reasonable to assume a measure of programmability in analytical sequences.

Consider an instrument programmed to make analyses on an increasing time base, such as

$$0, 5, 50, 100, 200, 400, 600, 800, \text{ etc. minutes}$$

for each sample, provided the parameter of interest has not exceeded a value somewhat lower than the agreed upon threshold.* The maximum intervals between samplings will have been chosen so as to ensure that the parameter cannot pass from this lower level through the threshold between successive intervals.

As soon as the parameter amplitude passes the lower level, the instrument changes to a more frequent time base. In the simplest case, it merely shifts to a shorter constant period, e.g. every 5 minutes. In a more sophisticated version, the period might be geared to the diminishing gap between observed and threshold amplitudes. In both cases, a measure of curve-fitting capability or graphic data display would allow the true incubation interval

---

* For example, the instrument may contain a second subordinate standard for this lower amplitude. The condition that this has been exceeded but not the main standard would set the instrument into its rapid sequence. The condition that both standards had been exceeded would terminate the analysis.

to be decided with precision sufficient for any practical purpose. (It is not at all impossible that after much experience with the shapes of such curves it might be possible to provide an *early* result by extrapolating to the instant of equality before the threshold has actually been reached.) Whatever the eventual form of the analysis, it will be obvious that precision can be maximized without incurring the inconvenience and expense of having to make every analysis implied by its level.

THE MULTIDIMENSIONAL DATA ENVELOPE

The time required for a food to spoil or otherwise become unwholesome (keepability), being the length of time it may be stored under given conditions before one or more parameters of unwholesomeness reach their agreed thresholds, can be determined from incubation intervals in a SMART analysis. Suppose we carried out many such analyses simultaneously on a homogeneous set of specimens, varying one aspect of the incubation conditions in a regular manner from specimen to specimen. For example, we might place the specimen containers in a temperature-gradient incubator for a maximum of, say, 100 hours. If we then plotted the observed incubation intervals

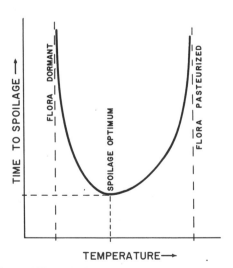

Figure 6-1. The keepability of a food specimen — as measured by the development of any single parameter of unwholesomeness — will be a function of temperature.

against temperature, a curve similar to Figure 6-1 might result.

Provided changes in the measured parameter resulted only from microbial activity and not from intrinsic processes in the food, the curve should tend asymptotically to the keepability axis both at very low temperatures (the microorganisms are dormant or frozen) and at some higher temperature (they have been pasteurized). A minimum keepability will be observed between these limits at the temperature most favorable to microbial activity.

In principle, we could carry out a still more ambitious experiment, varying two, three, or even more aspects of the incubation conditions in a similar manner. For example, we might vary pH,* humidity, partial pressures of oxygen or carbon dioxide, light intensity, and so on. Of course, the number of analyses will increase exponentially with the number of variables and quickly reach impractical proportions. However, varying two aspects, both of which limit microbial activity, might yield a three-dimensional surface similar to Figure 6-2. If one variable is not limiting (for example, oxygen partial pressure where all the organisms are facultatively anaerobic), a surface similar to Figure 6-3 might result.

To introduce more variables would lead to higher dimension surfaces, difficult to visualize and probably suffering from a shortage of data points but, in principle at least, handleable using computerized curve-matching techniques. These multidimensional envelopes completely describe a food's future behavior and would be, therefore, the most comprehensive form of microbiological data we could hope to produce.

In our present state of technology, such sophistication is probably out of the question. However, the lowest point of the envelope is unique and obviously relevant to absolute standards as it represents the minimum time for which the food will be wholesome, regardless of circumstances. There are situations where it may be worth determining since to do so should involve

---

* This applies to situations, for example, where the specimen under examination would normally be used (diluted) with another food. Thus, the buffering capacity of a chicken salad will affect the pH of a dressing mixed in with it to an extent determined by the relative concentration of dressing used.

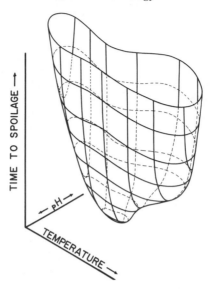

Figure 6-2. Schematic illustration of the keepability envelope for a food as a function of two variables where both variables (temperature and pH) may limit microbial processes.

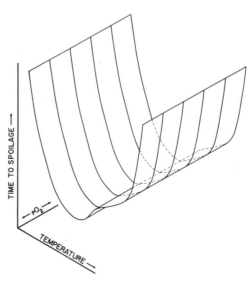

Figure 6-3. Schematic illustration of the keepability envelope for a food as a function of two variables (temperature and oxygen tension) where the organisms are relatively tolerant to oxygen.

considerably less effort and equipment than determining the whole envelope.

There are also situations where it may be valuable to know the general shapes of envelopes for different parameters, for example, when we wish to minimize analysis times by accelerating analyses. Foods are usually manufactured or stored so as to be as close to limiting conditions (temperature, pH, etc.) as possible in order to maximize keepability. Choosing a suitable set of graded conditions on the envelope, along which the curvature exhibits minimum variability, may allow keepability extrapolations to be made towards this limit with maximum confidence.

KEEPABILITY PREDICTIONS BY EXTRAPOLATION

While durations of compliance analyses in regulatory laboratories are always likely to be dictated by the incubation intervals specified in statutory standards, many situations call for more haste. In food manufacturing plants, for example, although it may generally be permissible to distribute produce without having microbiological data, early warnings of potential problems at retail outlets are valuable. Occasionally, it is necessary to hold produce, pending the availability of data; in such cases, rapid analyses are particularly desirable. The basic SMART methods of analysis may be appropriately foreshortened, albeit at the sacrifice of some precision.

Food of reasonable quality will only provide quick results if the analysis (which ideally should be carried out under exactly the same conditions experienced by the food) is short-circuited in some manner. For example, if the product is to be held at 5°C in the store, the ideal keepability analysis would also be carried out at this temperature. Unfortunately, the analysis would just be yielding a result by the time the food dies on the shelf. Obviously, the analysis must use a more propitious incubation temperature if it is to have any value.

Note that in such a circumstance, conventional plate counts should also be carried out at 5°C so as to enumerate only those organisms able to grow at this temperature. However, this also takes too long to perform so that plates are incubated at higher temperatures more favorable to growth. Unfortunately, many other organisms may then grow and are counted, though they might not have grown in the stored food. The practical, conventional analysis thus also short-cuts ideality at its own sacrifice of precision.

Before considering practical ways to speed analyses, we should make a most important comparison between conventional counts and analyses using parameters of unwholesomeness. It is reasonable to ask why, since the arguments in Chapter 5 emphasized precision or unequivocality as a major attraction of the approach, we can now consider the introduction of uncertainties into its data. Does this not reduce the value of the approach? I believe it does not.

First, the loss of precision under consideration refers only to SMART data, which are basically keepability data of mainly commercial significance. The precision of compliance analyses based on parameters of unwholesomeness is not in question.

Second, we must assume, in general, that analytical precision (or the certainty with which the datum can be reckoned to represent the true figure) and, therefore, the value of the result, will both decrease the greater the extent of the extrapolation and the number of variables involved. In the limit, when the extrapolation is greatest and the variables least gauged, we reach a value commensurate with that of the conventional count. Although it needs many hours of laboratory time to obtain, the count actually refers to the instant the specimen arrived in the laboratory. It is not usually referred to this way, but the necessary extrapolation implied in relating the count from a food to its keepability is a maximum.

Foreshortened SMART analyses, though lacking perfection, relate to a much later instant — the completion of the analysis — rather than to the time of arrival in the laboratory. The data also carry within them contributions from the relevant factors of microbial number, metabolic activity, multiplication rates, symbiosis or antagonism, and human sensitivities. Although the first four of these may be distorted by unnatural incubation conditions, they are, nevertheless, better measures than outright guesses on which to base extrapolations.

If we were unable to draw on experience, plate count data would be valueless, whereas accelerated SMART analyses would provide data that — though known to be wanting — could serve as bases for decisions. In the long run, we enrich plate count data with the value of accumulated experience. The SMART base is less flinty to begin with, but we can augment it in exactly the same way. It would be possible to develop statistical expressions describing, in general terms, relative variances for two techniques, one of which requires extrapolation over a longer interval and over more variables than the other. However, in the absence of actual comparative data to lend substance to the expressions, there seems little point in doing so. I believe the value of an accelerated SMART analysis will be greater than that of the corresponding plate count, but until data are available it is a matter for personal opinion.

With anticipating this criticism out of the way, we can return to the practicalities of minimizing SMART analysis times. There are at least two useful avenues. The first, involving minimal extrapolation, will result in minimal loss of precision but will be less generally applicable.

Extrapolation Along Amplitude/Time Curves

Experience may indicate that the amplitude of a parameter usually develops from a basal level of acceptability, through the physiological threshold in a smooth progression, yielding from the various specimens curves of characteristic shape and regularity. In QUANTI-TAINT, for example (see QUANTI-TAINT section), curves showing evolution of hydrogen sulfide by ground beef tend to change rather abruptly once they begin to show any inclination to do so, whereas those showing pH change in milk exhibit less sudden swings.

As soon as a definite amplitude change is detectable, therefore, we may be inclined to surmise the eventual course of its curve. In practice, and with accumulated experience, we might determine the shapes of the lower and upper 95 percent confidence limits for the relevant portion of the amplitude/time curve, as illustrated in Figure 6-4. Once a specimen shows data points inside these limits, the actual time to reach threshold may be predicted with the appropriate level of confidence.

The time saved will, of course, depend on the food, the normal rate of change of amplitude, and the level of confidence required in the result. If analysis conditions are identical to those

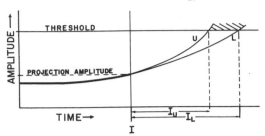

Figure 6-4. Curves (L and U) showing lower and upper 95 percent confidence limits may be determined for the growth of a parameter of unwholesomeness in a food, from a predetermined projection amplitude to the threshold of unwholesomeness. These confidence limits may be used to reduce the duration of future analyses as shown. In routine laboratory work, it would suffice to add values I_L and I_U to the measured interval I required for the specimen to obtain the shelf life.

experienced by the food, the saving may be relatively small, and the usefulness of the method will be rather limited. However, it may be valuable in at least two situations.

The first occurs when a quality control laboratory is geared to, say, an eight-hour working day in a plant operating longer or continuous hours. A proportion of the microbiological analyses undertaken will not be completed by the end of each laboratory day. There may thus be a sixteen-hour hiatus in information until the laboratory reopens next morning. In such cases, ability to anticipate incubation intervals by one or two hours may prevent the evening shift being deprived of the benefit of microbiological data.

The second situation occurs when raw materials themselves verge on being unwholesome. For example, the quality of raw meat shipments arriving at a factory vary greatly, but unless there are overt signs of putrefaction or other deterioration, impending microbiological problems cannot normally be detected. Since analysis speed in the parameter of unwholesomeness approach is related to the urgency of the situation, and if it is legally permissible to delay unloading shipments for a short time, it may often be feasible to detect incipient putrefaction before becoming committed to acceptance. Any ability to predict the course of a changing amplitude/time curve will obviously be valuable here. Except for interest's sake, it may not be too important to even know the actual time to threshold since any reliable change during the available analysis time will probably be sufficient grounds for rejecting the shipment. In this case, therefore, the microbiologist's confidence in the practical significance of an amplitude change may be more important.

Just how far one can consider taking such extrapolations depends on the limit of detection attainable for the parameter of interest. Figure 6-5 illustrates how an analytical procedure of lower noise level detects microbial activity earlier than one of poorer performance and may thus allow extrapolations to the physiological threshold over a greater interval. On the other hand, the resulting confidence limits will be wider. The trade is, as always, between speed and precision. Note that in the limit, this procedure reduces to the form of common nonnumerical analyses. For example, an analytical procedure of suitably low limit of detection may detect ammonia almost instantly in ground beef. The datum, however, pertains only to accumulated deamination activity in the meat; extrapolating from this to potential future activity invites confidence limits so wide as to be almost meaningless. The datum has, in fact, about the same value as the plate count as a predictor of future activity in the meat.

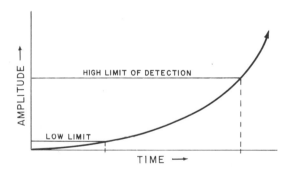

Figure 6-5. Analytical procedures with low limits of detection provide information on microbiological activity earlier than those with higher limits. In urgent situations, such prompt information is better than nothing. Note, however, that procedures with detection limits considerably lower than tolerable human response levels lead to very wide confidence limits if we use them to predict spoilage or hazard.

Extrapolations from Graded Incubation/Time Data

The following discussion is directed towards temperature as probably the most important variable on which extrapolations of keepability to normal storage conditions would be based. However, the argument applies to any variable.

Figure 6-1 schematically illustrated the keepability of a food sample as a function of temperature. If we took other sets of specimens in a similar manner but from different batches of food, we might obtain curves like I, II, and III in Figure 6-6. Obviously, curve III represents the sample of·highest microbiological quality since its time to unwholesomeness is greatest at all temperatures between the limits.

After examining a range of samples, we would probably observe that curves did not have identical shapes or show minima at exactly the same temperature. However, we could choose an arbitrary temperature — say, the most probable one for maximum deterioration rate — and superimpose all curves at this point. We could then draw lower and upper 95 percent confidence limits for the distribution of the remaining hours to unwholesomeness at temperatures on either side of this point, as shown in Figure 6-7.

This most rapid deterioration temperature then determines minimum analysis times. In principle, after determining the keepability of a food sample at this temperature, its residual life at any other, e.g. 5°C, can reasonably confidently be assumed to lie between the limits pertaining to that temperature. Its overall keepability would thus be expressed as a range, determined by the sum of the incubation interval at the worst possible storage temperature, plus the lower and upper ordinates for the confidence limit curve at the actual storage temperature.

The use of what is essentially this method has, in fact, recently been proposed by Hankin et al.[10] who observed that whereas initial bacterial counts on commercial pasteurized milk samples were of little value in predicting keeping quality, the number of days required for milk to go bad (organoleptic test) at one temperature was closely related to the length of time required for it to go bad at any other temperature. These workers suggested that practical rapid estimates of milk storage life at reasonable,

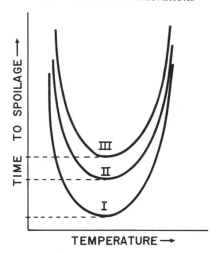

Figure 6-6. Each specimen of a food product will have its own keepability curve and its own minimum keepability.

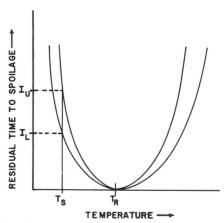

Figure 6-7. Confidence limits for residual keepability (or time to spoilage), formed by superimposing keepability data at an arbitrary reference temperature (T_R), will allow rapid prediction of spoilage times for other specimens at a realistic storage temperature (T_S). Intervals I_L and I_U are added to the incubation interval determined for a specimen at temperature T_R.

i.e. low, storage temperatures could easily be obtained from results of storage at higher temperatures.

In principle, it is possible to draw the curve of best fit to data from two, three, or more temperatures and from this make extrapolations to any desired temperature. Precision will increase with the number of temperatures studied, and the proximity of at least one to the actual storage temperature. In practice, the complexity (cost) of the equipment and technician time spent on each analysis will both rise with the number of temperatures studied to become limiting factors on precision. In addition, analysis times will increase as the actual storage temperature is approached, and this will reduce the value of the datum as a predictor. The individual situation must dictate where the compromise is made and thus the final form of the apparatus.

One futher possibility for reducing analysis times remains since the incubation interval data may themselves be extrapolated from amplitude/time curves in the manner described under the previous subheading. The same considerations of trading precision for speed will apply. Thus, analysis times may be shortened by pushing the limit of detection further and further below the physiological threshold; however, our confidence that the result reflects the value we wish to know will continually decrease. I believe that despite the loss involved, the potential exists for making relatively rapid analyses that, nevertheless, provide more immediately relevant information than plate counts and the like.

INDICATOR PARAMETERS

Whether or not one should consider carrying out analyses that merely indicate the likelihood of encountering some more malevolent presence is, I think, likely to be a more contentious issue in the parameter of unwholesomeness approach than it is in conventional microbiology. Classically, routine analysis for organisms indicating pollution by feces (itself being dangerous only when it contains pathogenic organisms) has great justification. Demonstrated presence of generally nonpathogenic organisms such as *E. coli* provides firm evidence of departure from good manufacturing practices, and because of their prevalence and higher concentrations in feces, detection is relatively simpler than for the more dangerous organisms. Data on indicator organisms can be a more significant contribution to the subsequent quality of food from a factory than the more direct alternative which, in all probability, will merely be a certification of the absence of overt pathogens in the specimen received.

Analogous analyses may be carried out in parameter of unwholesomeness procedures. Utilization of lactate, generation of formate dehydrogenase, and glutamate decarboxylase activities within a specimen, for example, could provide information about the ability of metabolic functions of *E. coli* or coliform organisms to increase in the specimen and, therefore, about the ability of a manufacturing process to avoid, injure, or destroy such contamination. The value of the information is very much open to personal interpretation, however.

On the one hand, if it has been adequately shown that, for example, formate dehydrogenase activity does not proliferate in the absence of fecal contamination, demonstrating its presence in a food provides just as valuable evidence of poor hygiene as the data from a plate or MPN count. On the other hand, formate dehydrogenase is not obviously a parameter of unwholesomeness, and there is little point trying to refer its level to any threshold of human response. How then, should we attempt to define recommended or statutory levels in foods? In any case, since formate dehydrogenase activity is not physiologically important to humans, would it not be better to improve its detectability by redistributing food specimens into selective artificial growth media similar to those used in conventional counting procedures? From this point of view, artificial growth media would certainly be acceptable, even preferable; however, such analytical procedures can barely be considered as belonging to the parameter of unwholesomeness approach. Might it not be reasonable to leave indicators to the monopoly of plate and MPN counts?

So far, I have tried to emphasize that although analyses based on parameters of unwholesomeness can be just as useful in the long run as microbial counts for monitoring and controlling manufacturing processes, they can also guard better against the consequences of statistical fluctuations in individual situations. Analyzing for indicator parameters rather takes us back to the "statistical blanket" level of information. The only justifications for considering such analyses further are that (a) they might be carried out on the same basic apparatus as true parameters of unwholesomeness and, (b) they would provide *some* information, and any information is better than none. My feeling is that, while it would be preferable to base parametric analyses and the certification of foods on its true hazards, indicators of fecal contamination are too valuable to overlook. The various possible chemical, biochemical, and immunological indicators should be included in the parameter of unwholesomeness approach.

Analyses for indicator organisms or parameters are closely related. Both seem to stand apart from other types of microbial analyses in that because they are more relevant to hygiene than human responses, their data should pertain to the instant the specimen was received. When we are not looking for the real pathogens, the quantity of feces entering food is most important, not its ability to proliferate thereafter. Parameter of unwholesomeness analyses only provide data about activity at the instant of sampling if the contamination level is already above the analytical limit of detection. A count, however, always relates to the instant of sampling; it may, therefore, be a better interpretation of the quantity of feces than the growth of an equivalent parameter. Whether or not it is in a given situation will depend on the limit of detection available in the biochemical analysis.

SYMPTOMATIC PARAMETERS OR MEASURES

Single, sequential, or continuous measurements of various physical, chemical, or biochemical parameters have been made on foods for many years. Much instrumentation has already been developed to that end. Many of the analyses fall within the framework of the approach described in this book, although only a few are used in exactly the same manner or with the same intention. All analyses provide useful information about microbial activities or presence, although the value of the information may vary considerably. Jay,[11] for example, has described forty-five alternatives to plate counts for assessing aspects of the microbiological quality of meats. Many are now obsolete or used very locally. However, many are potentially interesting as analytical bases in future instrumented methods. The majority do not measure true parameters of unwholesomeness; strictly we should call them indicators, but this term has a rather precise connotation in conventional microbiological analysis. Therefore, to avoid confusion in discussing their merits, I have grouped them together under this heading of symptomatic. They are parameters symptomatic of microbial contamination (but not necessarily fecal), and they are not necessarily parameters of unwholesomeness.

The outmoded "clot-on-boiling" test of milk quality, for example, actually assessed a milk spoilage parameter. In the sense of being a one-shot measurement with clotting tendency as the tolerance limit, it fits the parameter of unwholesomeness approach perfectly. On the other hand, it was not generally carried out sequentially during a prolonged incubation period so that precise grading of the better quality milk samples was not accomplished. In the same area, periodic inspection of incubated milk samples for accumulated reduction of methylene blue dye closely approaches the SMART method of analysis. Methylene blue is not, however, a parameter of unwholesomeness in that it is not directly relatable to any sensibility of the human body. The generally accepted objective of the test is simply to infer the concentration of microorganisms in milk without the labor of making plate counts.

Sequential or continuous determinations of the ability of extracted, incubated organisms to generate ATP, utilize ^{14}C-glucose, alter the optical density or electrical impedance and other characteristics of the growth medium, have all been instrumented in recent years and advertised with varying success as substitutes for viable counting methods in food microbiology. Unless selective growth conditions are used, these methods tend to simulate the standard plate count. Even so, abilities of the data to correlate with SPCs vary considerably for the reasons discussed in Chapter 4. Moreover, none of these parameters represent unwholesomeness as such. Nevertheless, because some instrumentation is already available and because future instruments developed for the parameter of unwholesomeness approach can easily include such symptomatic parameters in their repertoires, it would be unwise not to consider them along with the true parameters of unwholesomeness. As always, any information is better than none and if, for example, analyses for symptomatic parameters can yield data less expensively, more conveniently, or more rapidly than unwholesome ones, they may be extremely valuable. It depends very much on the situation, and microbiological analysis is never short of unique situations.

A most valuable use is probably in very rapid measurements of accumulated microbial activity at the instant of sampling. Brown and Childers,[12] for example, described the electrometric determination of Eh in ground beef. A simple electrode/meter system almost instantaneously determines its accumulated oxidation-reduction changes. A proportion of the reading is due to microbial activity. Intrinsic factors from the original muscle, preparational variations, and electrochemical variations between microbial flora contribute to a rather large apparent specimen to specimen variation when attempts are made to correlate the data with microbial numbers. This is quite the wrong thing to do, however. If we do not recognize the count as a very useful indicator of microbiological quality, Eh measurement is seen to be capable of grading ground beef specimens in its own right. It is potentially a very useful method, particularly when there is insufficient time or labor available to carry out more direct measurements of wholesomeness.

Its relevance to SMART analyses becomes very evident in further work reported by the same authors, where Eh of ground beef specimens incubated at 5°C, i.e. a typical storage temperature, was measured sequentially for thirteen days. In two days after Eh reached $-60mV$, samples rated as spoiled in organoleptic tests. If the objective of these sequential measurements had been to predict the onset of spoilage rather than merely to demonstrate the change of Eh with time, we would regard it as a SMART analysis (although, doubtless, the overall analysis time might be considerably reduced). Of itself, Eh is not a spoilage parameter, but it is symptomatic of the microbial growth resulting in spoilage. Neither its value or rate of change at any instant nor the time it takes to reach a prescribed value can be expected to correlate precisely with those of true spoilage parameters, such as the optical or rheometric properties, partial pressure of H_2S, etc., of ground beef. These are also unlikely to correlate with each other, of course. Eh is as valuable as a plate count; in sequential mode, it is likely to be a more reliable predictor of spoilage. In view of its ease of measurement, it should certainly be included in the repertoires of future instruments.

The situation is very similar for the classical microbiological oxidation-reduction indicators such as resazurin, methylene blue, and the tetrazoliums, which have been used extensively for milk and are finding increasing applications with meats.[13, 14, 15] While it may not be too clear just what determines the observed color of these substances at any instant, there is no doubt that the changes are loosely related to redox potentials in foods and to microbial activity. Resazurin has been favored for the rapid changes it exhibits — often within seconds of contacting a food. Its E_0 is high enough, however, to cause its reduction by intrinsic processes within many tissues. Because of this, it is unsuitable for prolonged incubations. Resazurin data may be regarded as reflecting something of the microbial activity accumulated up to the instant of sampling but no more. Methylene blue and the tetrazoliums require rather stronger electropotentials to drive their changes; they are more suited to demonstrating the ability of the flora to proliferate during lengthy incubations.

As with Eh, if microbial numbers are the objective, the data obtained from these substances do not correlate very convincingly with plate counts. On the other hand, when the data are regarded as expressions of integrated microbial activity and ability to proliferate, they will be seen to contain more information than mere microbial counts. The changes they display are not parameters of unwholesomeness and cannot be confidently related to human physiological responses; nevertheless, the analyses are so simple and so closely related that they should be considered in the main approach.

Radiometric determination of glucose or other metabolite utilization and electrical impedance changes must be appraised more carefully. Like Eh measurements, their data do not relate directly to the true parameters of unwholesomeness. In addition, these analyses usually require organisms to be transferred from the food into a medium more suited to the analytical procedure. In so doing, another rather important contribution to the total required information — the influence of the food itself on their fate — is lost. Such analyses are certainly interesting. They do at least provide some indication of whether or not the organisms are actually alive and, under favorable circumstances, reasonable correlations with, say, spoilage times, can occur. My feeling is that they should be regarded as forerunners of SMART techniques, though they will always remain on the fringes.

There are many other symptomatic parameters. Moreover, parameters describing unwholesomeness in one situation may not do so in another. Changing pH, for example, may be a parameter of unwholesomeness in the context of finished frankfurters but may only be symptomatic where the quality of some of its ingredients is concerned. Generally, if such analyses are easily incorporated into SMART procedures, we should do so but should always keep in mind that though the data may help in forming overall opinions of products or processes, they cannot unequivocally define quality. Only measurements based on parameters of unwholesomeness can do that.

DRY FOODS AND INGREDIENTS

The changing concentrations of microorganisms and the parameters by which we recognize them have so far been discussed as though they inevitably occurred within a food. How though, should we examine a material, if because it is dry or otherwise stabilized microbial growth and metabolism occur undetectably slowly, or if it is merely used as an ingredient in a finished food?

The former, for example, candy, biscuits, breakfast cereals, prunes, instant soups, etc., present little problem. If we are certain that microbial processes have halted, analysis for compliance with permitted amplitudes may be made as soon as the samples are received. Neither incubation nor sequential sampling are required; it should merely be necessary to homogenize or otherwise bring the food into a suitable state for analysis.

The latter requires more thought, however. Should one treat food ingredients such as egg or milk powders, thickeners, builders, extenders and gelling agents with the final composition of the food in mind, or should analyses be based on conditions more generally favorable to the onset of unwholesomeness? In some instances, the fate of the material is well defined. For example, alginate received at a factory may be intended only for an instant pudding mix; evaluation of its ability to generate parameters of unwholesomeness in incubated puddings — prepared from mixes identical to the normal manufacturing output — will provide the most realistic data. More rapid, though less useful, data could be obtained using conditions more favorable to microbial growth, avoiding low pH, preparing the mix at 50°C rather than 95°C, for example.

The main problem occurs with ingredients of virtually unlimited utility, such as retailed milk powder, flour, and spices. Conditions very favorable to the generation of undesirable parameters (such as infective doses of *Salmonella* organisms) will almost certainly occur among potential uses. It is not realistic to consider defining the quality of such products in terms of their utilization, e.g. standard A for the material used in baked goods; standard B when it is used in uncooked low acid foods, etc. Permitted levels will have to be defined with the most favorable circumstance for proliferation in mind and will, therefore, tend to be low. The most practical approach to analysis of these ingredients is probably the conventional microbiological procedure of incubating them in artificial growth media that are most favorable to growth of parameters of interest.

SAMPLING

The problem of obtaining suitably representative samples of foods for microbiological analysis and then of deciding how the data from them describe the overall quality of the lot from which they were taken is, logistically, one of the most formidable aspects of food microbiology. It surrounds but is rather separate from the less esoteric aspect of actually carrying out the analyses. The subject has recently been thoroughly covered through the auspices of the ICMSF.[16]

It would be presumptuous to claim that the parameter of unwholesomeness approach is likely to make great impact on the overall sampling problem. Nevertheless, through its potential effect on the level of instrumentation and throughput in microbiological analysis — by facilitating the analysis of larger numbers of specimens and possibly also the analysis of larger specimens themselves — it may alleviate some sampling difficulties.

A laboratory's choice of sampling plans — what it reckons to be a representative sample, the type of sampling scheme employed, and method of drawing samples — depends on considerations such as the degree of hazard involved, the uniformity, stratification, record of consistency, and other factors for the food, taken together with its practical limitations of analytical capability. Increased capability to analyze samples will obviously bear on the final choice of plans and may allow a more comprehensive sampling protocol to be established. Regulatory laboratories must also consider problems of collecting and handling field samples and units; in general, these are unlikely to be affected greatly by changes in the laboratory's capabilities.

The formal structures for acceptance or rejection decisions, which are based on the number of positives among a given number of samples, are known as attribute plans. The stringency of a plan is measured by the probability of accepting lots in which a particular proportion of sample units is defective. Attribute plans are not based on numbers of microorganisms per se but simply on a statistical manipulation of quantized compliance data, i.e. an individual specimen conforms to a standard or it does not. They are, therefore, independent of the analytical procedures used to obtain compliance data. Recognition of analytical methods based on parameters of unwholesomeness would thus not necessitate any modification of the acceptance/rejection principles that have been internationally agreed upon as desirable and that are currently in the process of adoption by several countries.

QUANTI-TAINT: A SMART INSTRUMENT

In order to investigate aspects of the parameter of unwholesomeness approach and so that we might give practical demonstrations of the form its instruments and techniques might take, colleagues Marcel Diotte and Istvan Dudas built the instrument now to be described. QUANTI-TAINT displays many of the features envisaged for future SMART instruments.

Figure 6-8. Two QUANTI-TAINT incubators with their microprocessor controller. Temperature control water baths and optional chart recorder not shown. QUANTI-TAINTS may be operated at different or identical temperatures, according to the analytical requirements and may also be operated independently. Multiple knobs (20) on each instrument allow control of sensitivity at individual specimen locations when QUANTI-TAINTS are used independently of the microprocessor controller.

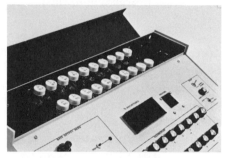

Figure 6-9. Showing the 20 specimen containers inside a QUANTI-TAINT.

It only differs, in fact, from the general instrument described in this and Chapter 5 in having a relatively simple sequential sampling system. Whereas the "all-purpose" SMART incubator is likely to have mechanical access to specimens so that portions may be physically removed to some of its detectors, QUANTI-TAINT features optical access only. Specimens remain undisturbed in their containers, and the instrument detects color changes occurring in the specimens themselves or in suitable indicators enclosed with them. Its scope, therefore, is more or less limited to detection of chemical parameters related to putridity, and from this it takes its name.

Figures 6-8 and 6-9 show the two QUANTI-TAINT incubators, each able to take twenty specimen containers, and the microprocessor controller. The incubators may both operate at the same temperature, in which case forty specimen positions

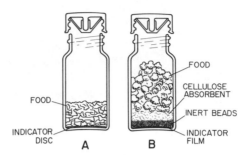

Figure 6-10. Typical QUANTI-TAINT specimen containers. A. Setup for detection of liquid-soluble metabolic products. Foods are pressed into contact with indicators absorbed in glass fiber discs. B. Setup for detection of gaseous metabolic products. Cellulose absorbent prevents water from weeping foods reaching indicator. Inert beads provide a uniform optical background for the indicator film.

are available. Alternatively, if the instrument is to be run in predictive mode, they will operate at different temperatures. Both incubators also contain scanning reflectance photometers, which sequentially inspect the transparent base of each specimen container at regular time intervals.

Figures 6-10A and 6-10B illustrate typical specimen containers and how foods are placed in them so as to enable either nonvolatile or volatile microbial parameters to be monitored. QUANTI-TAINT can provide data on, for example, pH changes within a food, evolution of hydrogen sulfide,* volatile acids, ammonia and other volatile bases, volatile reducing substances, the color of the food itself, and symptomatic parameters such as carbon dioxide evolution and reduction of resazurin or methylene blue dyes. Several analyses may be carried out simultaneously. Of course, since each analysis calls for one place in an incubator, increasing the number of analyses decreases the number of specimens that may be handled. Specimens may be added or changed at any time.

---

* At low vapor pressures for a gas partitioned between liquid (dissolved) and gaseous phases, a simple relation can be expected between its normal equilibrium vapor pressure and the rate at which it will diffuse into a reactive sink (the indicator). The early (increasing gradient) part of these curves is probably an accurate reflection of rising hydrogen sulfide vapor pressure. Later on, however, diffusion into the indicator probably becomes a rate-limiting step in the transfer.

If specimen containers have already been charged with indicator discs and/or reactive layers (Fig. 6-10B) foods may often be prepared for analysis simply by adding the correct amount to appropriate containers. For the methylene blue or resazurin reduction analysis, the quantity of food is not important, provided a minimum depth of 2 to 3mm covers the indicator disc. The atmosphere in the container must sometimes be replaced, using a rather simple apparatus developed for the purpose. For example, if the analysis simulates hydrogen sulfide generation in a gastight meat pack, air in the container must be replaced by nitrogen or carbon dioxide. Otherwise, evolution of the gas will not proceed properly.

QUANTI-TAINT is instructed with details of the specimen's origin and the level it should recognize as the threshold of unacceptability. Before inserting the specimen, the operator introduces a calibration standard that the instrument will read, remember, and subsequently use to compute the status of the specimen. At preset intervals, the instrument quickly (milliseconds) scans all specimen locations and stores reflectance values in its memory. Whenever the operator is not introducing or removing data, the controller screen automatically displays graphs showing the progress of specimens, one every fifteen seconds (Figure 6-11). The operator may call up individual graphs out of turn if he/she wishes. As soon as any specimen (or specimens) passes the predetermined threshold, an alarm sounds and the relevant graph(s) display, in turn, until the

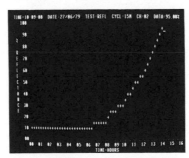

Figure 6-11. The QUANTI-TAINT microprocessor controller successively displays graphs showing the progress of specimens. As soon as any specimen violates its set threshold, an alarm sounds and the controller displays an alert.

operator acknowledges and acts on the data. In the meantime, data acquisition proceeds normally for the other specimens.

Figures 6-12 through 6-16 show typical changes occurring in ground beef and milk. They were recorded by Pearl Peterkin, using a pen recorder attached to the experimental instrument.† The graphs clearly show characteristic changes for the various parameters and how the QUANTI-TAINT incubation interval needed to generate a given change is shorter for specimens of decreasing microbiological quality. They also show how, for a given specimen, the incubation interval needed to produce a given change decreases as the incubation temperature increasingly favors microbial proliferation.

Lines representing thresholds of unwholesomeness or tolerances are also drawn on some of the graphs. Their intersections with the graphs indicate the incubation intervals that, in routine use, the instrument would either print out directly or use in calculating shelf life. These thresholds are purely hypothetical and are for illustrative purposes only;‡ practical values can only be determined from the exercises described earlier in the chapter. We have no immediate intention of determining amplitude/response data needed for the setting of these thresholds. It is sufficient at the moment to be confident that thresholds could be set and that analytical instruments having adequate performance to detect those levels exist or could be made.

The following examples show applications of QUANTI-TAINT data. Some of these are useful in their own right, but it should be kept in mind that much of the intended function of the instrument is to demonstrate the potential of parameter of unwholesomeness procedures. In other, more general instruments, data could be based on many parameters other than reflectance (viscosity, toxin concentrations, immunological or serological reactions, etc.). Thus, while it may seem rather pre-

---

† Pen recordings are shown here because of the poor vertical resolving power of the present controller display.

‡ Similarly, the useful analytical range of the instrumental procedure could be modified (increased or decreased) to accommodate practical thresholds outside the range shown.

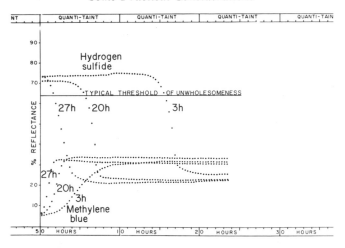

Figure 6-12. Sequential measurements of hydrogen sulfide generation and methylene blue reduction at 30°C, for ground beef specimens of differing microbiological status. Specimens had aged 3, 20, and 27 hours at 20°C before analysis in QUANTI-TAINT. Falling reflectance for hydrogen sulfide analyses indicates generation of the gas; shelf at low reflectance is due to depletion of the indicator. Rising reflectance for methylene blue analyses indicates reduction of the dye. An illustrative hydrogen sulfide tolerance level has been added to the chart. Note how residence times in QUANTI-TAINT before specimens reach agreed state of unwholesomeness provide measures of microbiological quality.

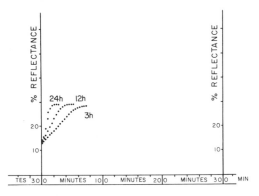

Figure 6-13. Sequential measurements of resazurin reduction at 30°C by ground beef. Specimens had aged 3, 12, and 24 hours at 20°C before analysis in QUANTI-TAINT. Rising reflectance indicates dye reduction. Owing to the rapidity of reduction, a fast chart speed and scan rate was used. Although the analysis grades many specimens according to an ongoing microbial activity, even sterile meat reduces resazurin rapidly, and analytical noise level is high.

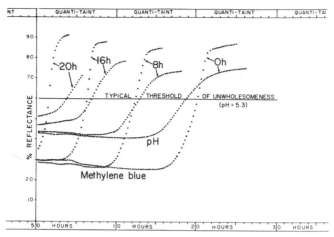

Figure 6-14. Sequential measurements of acid production (as indicated by bromocresol purple) and methylene blue reduction at 25°C for milk specimens of differing microbiological status. Specimens were inoculated uniformly with low level of souring organisms to ensure this course of deterioration; specimens had aged 0, 8, 16, and 20 hours at 20°C before analysis. An illustrative pH tolerance level has been added. Residence time in QUANTI-TAINT before specimens reach this agreed level of unwholesomeness provides measures of their microbiological quality.

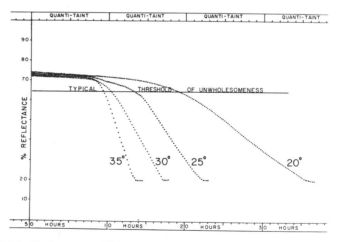

Figure 6-15. Hydrogen sulfide generation by (initially) identical ground beef specimens, recorded by QUANTI-TAINT at 20, 25, 30, and 35°C. Residence times before specimens reach unwholesomeness varies with temperature. Variation may be used to predict spoilage time at other temperatures. Slope variation of descending portions of reflectance/time curves probably results mainly from temperature dependent microbial activity, but hydrogen sulfide diffusion rate in specimen containers may also have an effect on slope.

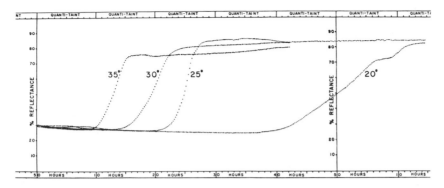

Figure 6-16. Methylene blue reduction by (initially) identical milk specimens, recorded by QUANTI-TAINT at 20, 25, 30, and 35°C. Residence times before appreciable reduction of the dye is visible vary with temperature.

tentious for QUANTI-TAINT to signal "HYDROGEN SULFIDE ALERT" for its ground beef samples, we should view this for its value in illustrating how an instrument using, say, radioimmunoassay based data might signal the existence of a much more tangible hazard.

Urgent Decisions

Situations where, for example, we need to decide very rapidly whether or not to accept raw material shipments and there are no overt signs of spoilage (odor, color, texture, etc.) do not call for demonstration of physiologically detectable parameter amplitudes. They require mainly that shipments having higher-than-normal microbial activities be quickly identified. The analytical procedures must have limits of detection well below those used for normal parameter of unwholesomeness determinations. Because indications of microbial activity must be sought at levels too low to cause responses in humans, all measured parameters must be considered as being only symptomatic.

The best possible limits of detection are required. In QUANTI-TAINT, hydrogen sulfide may be detected within thirty minutes in spoiled (twenty-seven-hour-old) ground beef (see Fig. 6-12), but some improvement is necessary before this, and ammonia evolution can be reliably detected long before spoilage is evident.

Resazurin reduction graphs for ground beef show increasing gradients as microbial activity accumulates (see Fig. 6-13). Intrinsic reducing activity in meat limits the detectability of microbial resazurin reduction; however, good and grossly contaminated specimens may be separable. Resazurin reduction data have conventionally been compared with plate counts and used as substitutes when the correlation appeared adequate. The parameter of unwholesomeness approach strictly calls for resazurin data to be related to more practical observations, such as the finished product's predicted shelf life, or postmarket performance. However, since these rapid decisions will not be based on physiological thresholds, there is no reason why a laboratory should not interpret resazurin reduction as simulated counts.

Visual inspection of pen recorder graphs from QUANTI-TAINT may be very effective in this situation. Alternatively, the microprocessor may be instructed to calculate reflectance change rates, say, every thirty or sixty seconds. Initially, these could be regarded as being related to microbial number. With increasing experience, they might be related to product quality. Either way, the microprocessor can be programmed to alert the operator quickly to signs of unacceptable levels of microbial activity.

Resazurin data are obtained very quickly. Many types of samples might be graded into, for example:

GOOD; CAN'T SAY; or POOR

within ten minutes. This is too short an interval for microbial multiplication to contribute significantly to the result. It does not, therefore, allow future activity to be predicted. The only claims that can be made for including resazurin and other symptomatic measurements (including parameters of unwholesomeness at levels below physiological thresholds) here are

1. They provide slightly more information than a plate count since they do measure an ongoing activity (although in the case of resazurin, the actual redox, enzymic, and ionic factors responsible are very poorly defined).
2. They are easily carried out with QUANTI-TAINT-like instruments.
3. Symptomatic information is better than none.

Compliance Analyses

In their simplest form, compliance analyses require parameter amplitudes to be measured once only — after the food has incubated under the conditions and for the interval defined in the standard. QUANTI-TAINT can do this. Its controller may be instructed to ignore a specimen container until a certain interval has passed. If the reflectance change then produced by the parameter in question exceeds its predetermined threshold, an audible alarm is activated, and a warning such as

SAMPLE 17 HYDROGEN SULFIDE ALERT ANALYTICAL LEVEL EXCEEDS RECOMMENDED THRESHOLD.

appears on the controller screen until the operator acknowledges the violation. If the specimen complies, QUANTI-TAINT records

SAMPLE 17 HYDROGEN SULFIDE — CLEAR

on the screen or in its printout.

While a single measurement may be convenient and preferable if the reagents for each analysis are rather expensive, or if the analysis itself takes considerable time, repeated analysis in QUANTI-TAINT costs nothing. The instrument will, therefore, normally read reflectance data from a specimen container at regular intervals, periodically displaying these as a graph on the controller screen. Graphs for the different specimens are displayed in turn until a violation occurs.

The only significant data here are ALERT and CLEAR. However, the sequential data available from the instrument are certainly interesting and valuable. Sequential sampling eliminates the possibility of false approval where, after an early violation, a parameter apparently decreases to an acceptable amplitude before completion of the standard incubation interval. In particular, it allows a warning to be given as soon as violation occurs, and this timesaving may be most desirable.

Monitoring Quality

It is here and in predicting time-to-spoilage that sequential sampling becomes essential and where only automated techniques can be considered practical.

For QUANTI-TAINT, incubation conditions are defined by the atmosphere (or other additives) in the specimen container and by the temperature of the incubator. Analytical precision — the number of grades possible — is determined by the maximum incubation interval allowed and the frequency with which specimens are inspected. Foods are then graded, as in Figure 5-4, on the basis of the incubation intervals necessary for parameters of unwholesomeness to reach agreed tolerances (thresholds of unacceptability).

Using conditions rather favorable to microbial proliferation provides an early result but one that must be extrapolated to normal storage conditions, as in Figure 6-7. Conditions similar to those of normal storage require longer incubations but lead to the maximum precision. The actual conditions chosen for any food will be a compromise between time allowable for the analysis, incubator capacity, and the acceptable confidence limits for the datum. In general, foods providing the most selective environment for microorganisms can be expected to exhibit the smallest confidence limits for a given extrapolation simply because the range of organisms capable of contributing to deterioration will be small. Such foods will, therefore, generally permit the most rapid analyses.

QUANTI-TAINT can store information about functions describing, say, residual time-to-spoilage versus temperature curves and their confidence limits. Having determined an incubation interval for a specimen, the microprocessor can then extrapolate the datum to any other temperature, displaying the result as the most probable interval for that storage temperature with its confidence limits.

When specimens can be satisfactorily subdivided for incubation at different temperatures, QUANTI-TAINT will be able to provide still firmer extrapolations by matching data to a curve with greater certainty. One temperature will usually be very favorable to the growth of a parameter; from this, the instrument will provide an early though rather uncertain indication of quality. The other will lie between this and the normal storage temperature. The second incubation interval will, therefore, be somewhat longer. After this further period, QUANTI-TAINT will provide a second, more reliable, estimate of specimen quality.

Beyond QUANTI-TAINT

From the previous discussion, it will be seen that both comparison of the microbiological quality of food against a standard and prediction of its condition at some future instant are technically feasible. Only one step — development of a sequential sampling incubator permitting actual mechanical access to specimens by whatever detectors are attached to it — stands between the limited capability of QUANTI-TAINT and instruments of the utmost versatility. I do not believe this step presents a great technical or financial problem, certainly not compared with that of developing realistic instruments to simulate plate counts. If the concept of defining food quality in terms of the true parameters of unwholesomeness rather than microbial numbers should be accepted, such future instruments will face few problems of credibility or verification. I believe the commercial attractiveness of such instruments will be very obvious. It needs only recognition of the concept by regulatory agencies to open the door to a flood of commercial instrumentation and automation developments in food microbiology.

REFERENCES

1. Jay, J. M.: Mechanism and detection of microbial spoilage in meats at low temperatures: a status report. *J Milk Food Technol, 35:*467, 1972.
2. American Society for Testing and Materials: *Basic Principles of Sensory Evaluation.* ASTM Special Technical Publication No. 433. Philadelphia, American Society for Testing and Materials, 1968.
3. American Society for Testing and Materials: *Manual of Sensory Testing Methods.* ASTM Special Technical Publication No. 434. Philadelphia, American Society for Testing and Materials, 1968.
4. American Society for Testing and Materials: *Correlation of Subjective-Objective Methods in the Study of Odors and Taste.* ASTM Special Technical Publication No. 440. Philadelphia, American Society for Testing and Materials, 1968.
5. Gilbert, R. J.: Staphylococcal food poisoning and botulism. *Postgraduate Med J, 50:*603, 1974.
6. Riemann, H. (Ed).: *Food-borne Infections and Intoxications.* New York, Acad. Pr., 1969.

7. Hauschild, A. H. W. and Thatcher, F. S.: Experimental food poisoning with heat-susceptible *Clostridium perfringens* Type A. *J Food Sci 32:*467, 1967.

8. D'Aoust, J. Y., Aris, B. J., Thisdele, P., Durante, A., Brisson, N., Dragon, D., Lachapelle, G., Johnston, M. and Laidley, R.: *Salmonella eastbourne* outbreak associated with chocolate. *Can Inst Food Sci Technol J, 8:*181, 1975.

9. Mossel, D. A. A.: *Microbiology of Foods: Occurrence, Prevention and Monitoring of Hazards and Deterioration.* University of Utrecht, Faculty of Veterinary Medicine, 1977, p. 8.

10. Hankin, L., Dillman, W. F. and Stephens, G. R.: Keeping quality of pasteurized milk for retail sale related to code date, storage temperature, and microbial counts. *J Food Protect., 40:*848, 1977.

11. Jay, J. M.: *Modern Food Microbiology.* New York, Van Nos, 1978, p. 125.

12. Brown, L. R. and Childers, G. W.: A rapid method of estimating total bacterial counts in ground beef. In *Mechanizing Microbiology,* (Eds.) A. N. Sharpe and D. S. Clark. Springfield, Thomas, 1978, p. 87.

13. Austin, B. L. and Thomas, B.: Dye reduction tests on meat products. *J Sci Food Agric, 23:*542, 1972.

14. Dodsworth, P. J and Kempton, A. G.: Rapid measurement of meat quality by resazurin reduction. II. Industrial application. *Can Inst Food Sci Technol J, 10:*158, 1977.

15. Bradshaw, N. J., Dyett, E. J. and Herschdoerfer, S. M.: Rapid bacteriological testing of cooked or cured meats, using a tetrazolium compound. *J Sci Food Agric 12:*341, 1961.

16. International Commission on Microbiological Specifications for Foods. *Microorganisms in Foods.* 2. Sampling for Microbiological Analysis: Principles and Specific Applications. U of Toronto Pr, 1974.

Chapter 7

ON THE ACCEPTABILITY OF METHODS IN MICROBIOLOGY

IMAGES OF MICROORGANISMS

. . . almost any germ in the environment can cause trouble. . . .

. . . allowing the tiny germs to multiply like crazy. There's nothing that shouldn't be watched. . . .

. . . can multiply to more than 1,600 million in eight hours. . . .

. . . health department warned on the weekend 2,500 cans of poisonous mushrooms which have been distributed. . . .

NO POISONOUS MUSHROOMS IN OTTAWA HOSPITALS

YOUR LUNCH: IS IT POISON?

LEAVE POISONOUS BACTERIA OFF YOUR NEXT PICNIC GUEST LIST

Every year about 400,000 Canadians are poisoned — without knowing it. The source of this malaise is the food they eat and the organisms it carries. . . .

. . . hamburger contaminated with poisonous organisms. . . .

How did the lucky 50 manage to escape? It's a cinch they didn't duck out for a hamburger — not with the bacteria levels in ground beef. . . .

FOOD POISONING FELLS 150 ON JAPANESE JET

. . . about 40 residents and 25 employees were felled.

. . . could create 'bacterial monsters.'

. . . says one package of bologna, ham . . . may contain up to 12,000 times more bacteria than the competitive package next to it.

. . . tests found bacteria counts up to four million per gram in cooked food.

. . . a $1.25 sausage plate . . . came second with 1.8 million bacteria. . . .

. . . hot dog, 40 cents, 750,000 bacteria. . . .

DANGEROUS GERMS POLLUTE EXHIBITION FOOD

. . . hamburger contaminated with poisonous organisms. . . .

. . . 49 [cheeses] were found to contain . . . disease producing strains of coliform organisms. . . .

THESE RATHER eye-catching phrases were taken from widely read daily newspapers. Whether they accurately portray the facts and whether they were penned by the various journalists out of concern for public well-being is irrelevant here. What is important is that they were written by persons successfully trained in a profession manifestly capable of influencing human motivations. The information in the phrases is presented in dramatic and emotive wraps — in attention snatching terms, powerful because they excite subconscious desires or fears.

The theme of this chapter is the suggestion that would-be developers of new microbiological methods should pay at least as much attention to the psychological impacts of their developments as to their potential scientific value. Many widely used scientific methods have not necessarily achieved the popularity they enjoy on the grounds of "cold, detached, scientific logic." Enumerative microbiology is one of these, for its efficiency at controlling the wholesomeness of food may be challenged, as was seen in Chapters 1 and 4. If its superiority can be questioned, we should examine why it has not long been supplanted by more informative procedures; finding a satisfactory answer is important to the future. The panegyrics of journalists may not at first sight appear very relevant to this, but in discussing the idea that microbiologists — just like any other scientists — are much more human than they are generally given credit for, the relevance will become obvious.

Scientists are not impartial or unbiased. They are as ambitious and have as many axes to grind as anyone else; science would stagnate if they did not. Statisticians, in particular, have long been aware that scientists and technicians must not be allowed to know (with the best intentions in the world) the relation the observation they are making has to the probable outcome of the experiment. Many nonscientists have a very distorted image of science and scientists. Terms like "cold scientific logic" and "scientific impartiality" are themselves only journalistic fabrications — quasi-religious phrases, popular and effective because they satisfy subconscious hungers for the existence of superresponsible beings.

It is also very relevant that microorganisms are not like pesticides and many other modern problems; they are, as it were, an archetypal enemy. Scientific methods for detecting them have been available for many years, and many of the concepts relating to them permeate family and social behaviors. Most persons in developed countries are overtly conditioned from childhood to associate microorganisms with germs and fear and to respond to situations involving microbes with avoidance patterns of behavior. Conditioning is received, for example, through journalese such as the expressions at the beginning of this chapter or with exhortations to wash hands, return dirty cutlery, or use antimicrobial substances in the most (scientifically) improbable places. We are even instilled with anthropomorphized interpretations of microbes revelling in the warmth or snoozing in the cold by regulatory and other food handling/safety organizations. An advertiser's *story board* used in the direction of TV advertisements, such as the Lysol Spray one shown in Figure 7-1, is a perfect example of conditioning, calling as it does for fear-reinforcement props like distorted electronic sound effects and

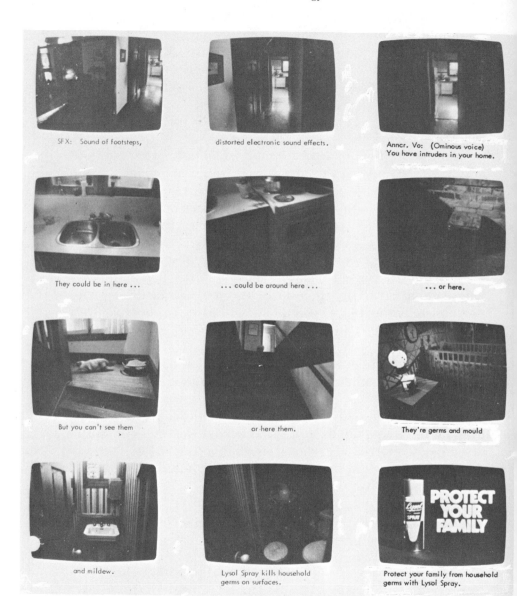

SFX: Sound of footsteps,

distorted electronic sound effects.

Anncr. Vo: (Ominous voice)
You have intruders in your home.

They could be in here ...

... could be around here ...

... or here.

But you can't see them

or here them.

They're germs and mould

and mildew.

Lysol Spray kills household
germs on surfaces.

Protect your family from household
germs with Lysol Spray.

Figure 7-1. An advertising storyboard used in the direction of TV commercials. Reproduced by courtesy of Sterling Drug Ltd., Aurora, Ontario. The spelling error in frame 8 was caused by a human foible.

the announcer's ominous voice. Once upon a time we distributed charms around ourselves to ward off evil spirits. Today, the same unreasoned fear of things unseen helps to sell mass-produced charms.

The subconscious influence of microorganisms has still more primitive roots, however. Long before the concept of microorganisms became fashionable, *Homo sapiens* had evolved organoleptic responses to provide warnings of many hazards, predominantly microbial in nature. Almost certainly, for example, our acute sensitivity to and distaste for sulfides and similar odors developed, not as a means of avoiding the chemicals themselves but rather the likely consequences of ingesting anaerobic microbial growths. We, therefore, carry deep within us (laymen and scientists alike) instinctive responses to many situations involving microbial contamination. Anyone in doubt of this should merely read through the following list of words and phrases, picked in less than two hours from *Roget's Thesaurus*. They are all pertinent to the possible existence of a microbial hazard, although most of them predate the concept of microorganisms by many generations, and the underlying relationship is really only the hint of danger from the intangible.

abhorrent	baleful	cheapened
abominable	beastly	cloacal
abscess	beslimed	clotted
abused	bitter	coarsened
acetous	black	collied
acrid	blighted	communicable
addled	blotchy	condemned
adulterated	bogey	consumed
affected	boggy	contagious
afflicted	bored	contaminated
ageing	botchy	corroded
allergenic	brackish	corrosive
ammoniacal	broken-down	corrupted
assafoetid	caked	crabbed
astringent	cancerous	crawlies
atrophied	cankerous	crawling
awful	carious	crippling
bacteria	carrion	crumbled
bad	catching	cursed
bad eggs	center of infection	damaged

damnable
dangerous
deadly
deathly
debased
decayed
decomposed
decrepit
defaced
defiled
deformed
denatured
deplorable
desecrated
detestable
detrimental
devalued
diabolic
dire
dirty
disabling
disagreeable
disastrous
discharging
discolored
disgusting
disintegrated
disordered
disrupted
distasteful
distorted
distressed
dreadful
dreggy
dross
dungy
eaten away
ebbing
eczematous
envenomed
eroded
cvil
excrement
execrable
exhaust
exposure

faded
failing
fatal
faulty
fecal
feculent
fell
fermented
fetid
filthy
flat
flawed
flyblown
foul
frowsty
fulsome
fungus
fusty
gamy
gangrenous
garbage
gassy
gathered
germ laden
germs
ghastly
gnawed
goaty
grate
graveolent
grievous
grisly
gross
grotesque
gruesome
harmful
hateful
hazardous
heady
hideous
high
hogwash
horrendous
horrible
horrid
hurtful

ill
ill-scented
impaired
impetiginous
incurable
indecent
indigestible
infested
injurious
in ruins
insalubrious
insanitary
insecurity
insidious
insulting
intolerable
irredeemable
jarred
jeopardizing
junk
lapsed
latrine
leathery
leprous
loathsome
lousy
low grade
low quality
maggoty
make one's flesh creep
make one shudder/vomit, etc.
malevolent
malignant
malodorous
mangled
marasmic
matted
matter
mattery
measly
menace
mephitic
messed
miasmal
microbes
mildewed

mired
mischievous
miserable
monstrous
morbid
mote
mothery
mouldered
mouldy
mucinous
mucky
mucus
musty
mutilated
nasty
nauseating
necrotic
nidorous
niffy
no good
noisome
nonsterile
noxious
nutty
objectionable
obnoxious
obscene
odious
offensive
ominous
on downgrade
outrageous
overpowering
past its best
pathogenic
pathological
pediculous
peril
perished
perverted
pestilent
piggy
pigswill
plague ridden
plaguy
poisonous

polluted
poor
profane
purulent
pus
pussy
pustulent
putrescent
putrid
putrifying
rancid
rank
ravaged
reasty
reeking
repellant
repugnant
revolting
ringworm
risky
ropy
rotten
ruined
running sores
scabby
scaly
scandalous
scatheful
scorbutic
scoured
scrofulous
scummy
scurfy
senescent
septic
serious
sewer
shameful
sharp
shocking
shoddy
shotten
shrivelled
sickening
sinister
sink

sliding downhill
slimy
slithering
slobbery
smell like a drain
smell like death
smelly
smut
snot
soiled
sordid
sour
spoiled
spotty
squalid
stained
stale
stench
stick in one's throat
stinking
strong taste
sullied
sulfurous
suppurating
susceptible
swollen
tabid
tainted
tarnished
tart
thick
threatening
toadlike
toxic
trash
trauma
treacherous
turned
turn one's stomach
twisted
ugly
ulcerated
unappetizing
unattractive
unclean
undermined

undesirable	upset	weeping
undrinkable	vapid	whiffy
uneatable	venom	wilted
unfresh	verminous	withered
unhallowed	vicious	woeful
unhealthy	vile	worm-eaten
unholy	virulent	wormy
unhygienic	virus	worn out
unloveable	vitiated	worse for wear
unpalatable	waning	worthless
unsafe	warped	wrecked
unsightly	warty	wretched
unsound	wasted	wrinkled
unwholesome	weakened	wrong taste

Languages only become so rich in related expressions when the subjects reflect very basic human motivations like eating, religion, or fear. It might be an interesting and informative exercise to compare the size of this list with one made up from terms pertinent to, say, laser ranging, or Raman spectroscopy. Unlike microbiology, many scientific subjects do not have a deep associational content; they are significant only at the conscious level.

We are entering a particularly sensitive and subjective area, where dissenting pokes at the concept of scientific rationality are likely to stimulate equally opinioned responses. Nevertheless, the possibility of the popularity of enumerative microbiology being maintained by a crutch of nonrational psychological factors should be discussed. Certainly, from what has been said, if there is *any* possibility of instinctive or conditioned responses influencing the acceptance of scientific methods, their effects are likely to be most pronounced in the area of food microbiology. Developing new methods is a reasonably respectable occupation and scientific merit a justifiable objective, but it is hardly satisfying to get to that point alone. The successful popularization of any radically new approach to food microbiology may depend not only on its ability to challenge an overwhelmingly greater body of scientific data and experience built up over the years by the existing approaches but also its ability to exploit the emotional responses of scientists, laymen, lawyers, and politicians alike.

IMAGES OF OBJECTIVITY

The acceptability of nonenumerative methods of analysis in food microbiology is low. Over the years, many alternative and often intrinsically more easily instrumentable methods of assessing the microbiological status of foods have been examined, without becoming widely accepted and used — methods based on, for example, pH, dye reduction, reaction with thiobarbituric acid, evolution of ammonia or volatile reducing substances, and changes in the various physical properties of foods. In fact, even the acceptance of modifications to accepted methods is slow. Compared with it, the prognosis for a new method of detecting Fenitrothion,* for example (which will enter a much less experienced and more flexible environment), will be much more encouraging.

The scientific explanation for the rejection of any alternative microbiological method has usually been that its data did not correlate with those from enumerative methods. This is a rather one-sided conclusion, however, savoring more than a little of geocentrism. On occasions, the prevalent bacterial chauvinism has even led to reversals of logic and orientation such as

> A disadvantage of organoleptic measurements is, however, that changes which are observed may not be directly related to numbers or types of bacteria present.[1]
>
> With respect to the sensory evaluation of meat freshness, a previous study . . . showed that tactile response correlated better with bacterial numbers than either odor or color. . . .[2]
>
> Organoleptic properties of ground beef . . . were studied in relation to the objective tests of total bacterial counts. . . .[3]
>
> Although the panel did not highly relate extract release volume to the organoleptic qualities, this test correlated four of six times to bacterial numbers . . . ERV determinations would be a much faster way of indicating the approximate quality of beef. . . .[4]

In other words, never mind the quality, feel the width. Bearing in mind the shortcomings of data based on numbers of microorganisms and the fact that microbial metabolic activity data may often be argued to be more directly relevant to wholesomeness

---

* O,O-dimethyl O-4-nitro-m-tolyl phosphorothioate, used for control of spruce budworm.

than are numbers alone (see Chapter 4), the correctness of such conclusions may seem open to challenge.

In fact, it is scientifically quite unjustifiable to arbitrarily assess one set of data in the light of another, even though the latter may be long accepted. When correlations do not occur, *both* data bases should be held up and evaluated for the needs of the moment. I believe there are many times when this has not happened in food microbiology. If this is so, the failure of alternative approaches to supplant enumeration must be ascribed to one of two causes:

1. their data *are* manifestly inferior at describing the microbiological status of foods, or
2. data couched in terms of numbers of microorganisms are more acceptable, i.e. believable, to many scientists than the value of the information in them actually warrants.

Can (2) be so? Is it reasonable to suggest that the idea of enumerating microorganisms in food has sufficient subconscious appeal to disturb the objectivity of scientific judgments? Proving or disproving this could be difficult, if not impossible, and I shall not attempt it. However, I shall try to peck a little at the crust of impartial scientific judgment and leave whatever may be exposed to the reader's interpretation.

Piltdown

The very human motivation of eminent scientists duped for thirty-seven years by the fake "Piltdown skull" is an excellent case in point. The "find" was first published in 1912. From it Teilhard de Chardin first gained fame, and from it Arthur Smith Woodward, Arthur Keith, and Grafton Elliot Smith eventually derived two Sir Arthurships and a Sir Grafton. Timings of both the first find — at a most opportune moment during the development of anthropological arguments — and of subsequent finds at the perfect instants to dispel detractors were themselves remarkably providential. Protagonists of Piltdown ignored dental and anatomical evidence and even crude file marks on its teeth in their eagerness for evidence fitting preconceived ideas. French anthropologists, blessed with an abundance of Neanderthal and Cro-Magnon relics, had for years rubbed English noses at their apparent lack of ancestry. Piltdown now provided Englishers with their own ancestor, even older and, moreover, with a fully modern braincase. When compared with progenitors of nonwhite, non-European races, Piltdown buttressed white supremacy. Gould[5] has written

> We know in retrospect that Piltdown had a human cranium and an ape's jaw. As such, it provides an ideal opportunity for testing what scientists do when faced with an uncomfortable anomaly. . . . If we are to learn anything about the nature of scientific inquiry from Piltdown — rather than just revelling in the joys of gossip — we will have to resolve the paradox of its easy acceptance. I think I can identify at least four categories of reasons for the ready welcome accorded to such a misfit by all the greatest English palaeontologists. All four contravene the usual mythology about scientific practice — that facts are "hard" and primary and that scientific understanding increases by patient collection and fitting together of these objective bits of pure information. Instead, they display science as a human activity, motivated by hope, cultural prejudice, and the pursuit of glory, yet stumbling in its erratic path toward a better understanding of nature.

APPERCEPTION AND MOTIVATION

The psychological processes that go beyond actually seeing an object (or concept) and involve recognition, understanding, and interpretation are known as *apperception.* This is the total process, suffusing the object with intellectual, emotional, and associative meanings. To the science of advertising, apperception is far more important than mere perception. What the mind does with sensory data is far more important than the actual data received by the perceiver; the same data may be apperceived (without fault) in different ways by persons of different backgrounds and outlooks. The mind's apperception of an object, the intangible interpretation it carries with it, is known as an *image.*

It is a well-established fact that human motivations can often be manipulated by the creation of images that stimulate instinctive or conditioned responses. The laws of and the possibilities relating to this manipulation are vitally important to politics, advertising, and other mass persuasion industries. Images with strong emotional associations are likely to have the most powerful motivating effect. Thus, for most adults, advertising exploiting a variety of subconscious fears can successfully stimulate nonobjective purchasing behavior for antimicrobial products. In contrast, products with weaker associational links tend to be viewed much more objectively. Few children, for example, are exposed to bogeys in the guise of flame ionization detectors, rotating viscometers, or arbor presses and if they purchase these in later life are likely to be less affected by the wording of the advertisement than they would be for deodorants or diet aids.

Consider a typical example of how the image created by information may affect its impact:

A research laboratory observes the oxidation-reduction potential of sausage meat to decrease with the age of the meat, while the total aerobic bacterial count rises to a plateau. In a carefully controlled experiment, sausage samples having oxidation-reduction potentials lower than -70mV are found to have a 75 percent probability of yielding plate counts greater than 10^8 per gram. Three different laymen (A, B, and C) are now presented with information about the quality of their lunches, phrased in different (but equivalent) ways. They are told that the sausage they are eating

A. contains one hundred million bacteria per gram
B. has an oxidation-reduction potential of minus seventy millivolts
C. has an oxidation-reduction potential of minus seventy millivolts, which could mean it contains one hundred million bacteria per gram

Layman A may throw down the sausage in horror. The impact of the statement is strong because childhood conditioning causes him to associate *bacteria* with *germs, uncleanliness,* and *fear.* Amplifying the apperceived fear is the association of number with strength. At one hundred million, the enemy are very strong — there is great danger.

B continues eating. If oxidation-reduction potential means anything at all to him, it is unlikely to be associated with danger, for he has not received that conditioning. The quoted number is not startling, although he may just wonder whether he should be experiencing a mild electric shock. All the associative links are weak, and the impact is ineffectual.

C, like A, apperceives danger in the statement about bacterial numbers. But a doubt is introduced regarding its true magnitude. This is his first meal of the day, for he has been down on his luck. He rightly resumes eating; fear of starvation is more real than the one hinted at. (The bacteria were harmless anyway, merely improving the nutritive value and flavor of the sausage.) Though the count may naggingly persist, the other datum will soon slip from his mind. The impact of this statement was tempered by conflict with another fear and particularly by the suggestion of unreliability.

Expressions like *oxidation-reduction potential,* along with *micromolality, hours to reduce,* and *percent transmittance* are typical of the data terms obtained from alternative methods of microbiological analysis. They have little associational content. It is probably easy to agree that for laymen at least, the psychological impact of such terms is very ineffective compared with conventional numerical data.

But what about scientists? Microbiologists receive the same conditioning as other laymen long before their microbiological training. Can we really assume that their perceptions are uncolored by it, simply because they have also learned some other facts about microorganisms? Does formal training condition out associational responses to microbial numbers?

In fact, it is possible that formal training even reinforces early conditioning by

a. emphasizing the tangibility of microbes — their "individuality" as colonies on agar, or their various likes and dislikes in regard to living conditions.
b. teaching receptive undergraduates the "normality" of the number concept in standards, guidelines, and official methods.

A recent, rather pertinent comment on both conditioning and scientific motivation[5] began, "Viruses are beginning to be baptised with latinised family and generic names, in the style of Linnaeus. Why? Is this artificial marshalling of sub-organisms necessary or useful? Or does it reflect virologists strict potty training?" Later, however, it suggests that ". . . the main and real reason that animal virologists want latinised names is because doctors treat people and (as with priests, politicians, administrators and lawyers) they have found it useful to develop a profound-seeming language, to maintain distance from and height above the people. So they appear more learned, safer from challenge, and can collect more fees (or souls, or votes, or red tape). It is helpful to invent confusing and impressive names for viruses and to shun the common ones like mumps, measles, chicken pox or yellow fever. Mumps and measles are undignified. . . ."

The normal formal training in microbiology may thus produce a mental set tending to exclude any future concepts that, as it were, de-emphasize the microorganisms. Whether this actually happens is eminently arguable, of course, just as any conclusions drawn on it are likely to be delightfully subjective. For the moment, I prefer to think of most scientists — microbiologists or otherwise — as not immune to the workings of the subconscious. But it is a matter of opinion.

Let us pursue the train of thought from the psychological impact of data expressions to the acceptability of methods and instruments. It is a fact that most inventions and developments in food microbiology turn out to be less than world shattering. Do those which fail to appear on every laboratory bench (though they might well have been useful in their own ways) do so because more factors than scientific merit are important to success? And might this be because in food microbiology far more than in most areas of science would-be developers (myself included) have assigned too much importance to the scientific or economic merits of their developments, believing that these were the grounds on which they would be accepted — wrongly believing, that is, that this was what potential users really wanted?

I now believe it is not necessary for one instrument or method to be demonstrably superior to others in order to be preferred. Underneath any of the modern veneers — civilization, science, or what have you — the average microbiologist (or any other scientist) is basically little different from the average purchaser of a loaf of bread or an automobile. While they may believe they choose a certain brand on the basis of a certain line of reasoning, they may be motivated by forces and processes they are not necessarily aware of. The scientist may feel he reaches the conclusion to accept a particular experimental method by logical analysis, but if a choice of equivalent methods is available, could his conclusion be dictated by the relative subconscious images they project?

And what significance would this have to those scientists developing alternative microbiological methods? Could spending a few months exploring the effects of motivations on the acceptability of scientific methods, instruments, or data be more rewarding than years of development work?

Certainly methods, designs, and terminology are rarely so inflexible that they cannot be modified to improve their impact. Had the previously mentioned research laboratory, for example, paid attention to the phrasing of its data, a statement such as

The sausage you are eating . . . is in a state of incipient putrefaction. . . .

might have stimulated layman B to react very much like layman A, for its emotive impact is much greater than an expression of oxidation-reduction potential and may be as great (if not greater) than an expression of bacterial number.

One may argue that a microbiologist possessing knowledge of the relations between oxidation-reduction potentials, bacterial numbers, and the onset of putrefaction in sausage is less likely to be moved by the change in terminology. But is he? Faced with a choice of equivalent methods, one providing data instantly perceivable by all in terms of unwholesomeness or disease, the other only after conscious manipulation, which method will he choose? I think that if his overall motivations are towards honesty and sympathy with the public, and if the data expressions were chosen carefully enough to avoid hints of showmanship, he will choose the former. On the other hand, of course, desires for aggrandizement and professional exclusivity could orient his preference towards data protected by quasi-mystic jargon, as one sees in medical terminology. There are all kinds of scientists, just as there are all kinds of people.

To be accepted as providing advantages over those established, a new method not only has to provide the advantages, it has to be perceived (or rather, apperceived) to do so. Statistical correlations may adequately illuminate the way when the choices are, say, methods of connecting optical fibers in telephone links. For a subject like food microbiology, however, which is close to the dark corridors of mind and instinct, the statistician's guiding light may glow dimly against nature's. The statistician's concepts of correlation, reliability, and so on are different and much colder things than the human feelings (images, apperceptions) on which acceptance is likely to be based. For example, while we may be happy with the claim that a method correlates 80 percent of the time with standard plate count results, we still have doubts and fears when it is *our* data that are in question. (Are 80% of my data correct, or am I one of 20% of people whose data is never correct?) Pascal expressed it rather nicely: "If the greatest philosopher in the world should find himself upon a plank, wider than actually necessary but hanging over a precipice, his imagination will prevail though his reason convince him of his safety." Because the importance of the data increase as we get close to a statutory or recommended limit, many samples suspected of being over the limit may have to be repeated using the established method — which rather detracts from the value perceived in the new method.

Acceptance or rejection of a scientific concept may occur formally, as when, for example, it is written (or otherwise) into a recommended procedure or at the individual level. An individual may mentally accept or reject a concept — in the sense of being positively or negatively oriented to it — while still conforming to a formal dictate. Images of the concept will answer or obstruct subconscious urges and thereby bias arguments used for or against its use. We never have *all* of the facts; because of this, it is quite possible for opposed arguments to be equally valid. The following examples illustrate subconscious persuasions influencing not only decisions but also the arguments used in their justification. They graduate from a domestic setting, where human influences may be more acceptable, to the laboratory:

In the stores, one brand of bread — a loaf of equal or better quality — was failing miserably against its competitor. Careful analysis indicated that consumer reaction was entirely turned by the precision with which the loaves were formed. The first loaf, having excellently formed rectangular sides, was found to project an image of machinery — of unnaturalness and interference — whereas its competitor, poorly formed and bulging with an exuberant lack of control, projected an image of natural wholesomeness. These associations, reinforced by the careful conditioning we receive to equate naturalness with superiority, completely dominated consumer preference. Conscious analysis of the relative values of the two products would probably have led purchasers to choose the first loaf, for it might have made neater sandwiches or fitted the toaster better. Preference was dictated instead by subconscious associations. The example was taken from a textbook of advertising psychology.[6]

In this example, the purchaser did not have to justify his/her choice to anyone. Consider now a discussion between a young couple buying an automobile. The husband's arguments may run something like

"A bigger vehicle would be so much safer in an accident. We should have the biggest engine. One day the extra power may get us out of trouble. Maybe even save our lives.

"With the big engine, we'll save more on repair bills than we'd spend on gas.
"We need room to put 4 × 8 panels. Remember the problems we had when
we built the rec room last year.
"With the state of the roads these days a good ground clearance is essential."

He is getting around to arguing for a truck. His arguments are
quite valid, though he could just have convincingly presented a
completely opposed set. His wife may do just that:

"A smaller vehicle is so much more maneuverable, we'd be less likely to be in
an accident.
"The mileage we do, we won't be expecting big repairs anyway.
"For just once in a year, we could use a roof rack.
"Smaller cars can drive around potholes that would catch a bigger vehicle."

Her arguments are equally valid, though she is after a com-
pact. The same information is available to both, but the importance each attaches to the various items differs. Both could find
national statistics or local examples to support their views. (The
credibility of either side will depend on the reader's own persuasions.) The information apperceived by the husband convinces
him that his wife's arguments are erroneous. Only a truck, not a
Volkswagen Rabbit, satisfies his image of the essential vehicle. Of
course, the truck image excites his subconscious desire to appear
masculine and self-sufficient, although no doubt even the
"macho" image itself was carefully cultured in his subconscious
by other men in downtown agency offices.

The microbiologist has been called in to discuss control of Algisnack, the
company's new product. He proposes following the fortunes of microorganisms using the plate count method. However, the product manager,
knowing they are simpler and cheaper to perform, suggests he uses the
methylene blue reduction (MBR) time method instead. The microbiologist
argues that "The MBR test is used hardly anywhere now because its data do not
correlate with plate counts. It is too difficult to interpret because different
organisms have very different rates of multiplication and metabolic activities."
To which, the product manager may counter, "But surely the plate count
doesn't tell you the metabolic activities of the organisms you count. Isn't the
MBR result more relevant to total metabolic activity and therefore to the
quality of Algisnack than the plate count?"

The product manager may feel the microbiologist has already
thrown a stone against his own argument. The microbiologist
sees that the manager is oversimplifying the problem. Both views
are probably valid, to a point. They will probably settle for the
plate count, for the microbiologist should easily overwhelm the
others in any microbiological argument. If he were given to

introspection, however, might he admit to taking a stand on grounds of a more human origin, not because the plate count unequivocally gives a more accurate picture but through fear of being a lone sheep? Suppose other microbiologists use the plate count for similar products, the Association of Official Analytical Chemists' recommended method is the plate count, and (in particular) the law dictates that the bacterial count in this type of product shall not exceed so many per gram. Adhering to the accepted method provides him with a personal insurance should a serious contamination problem ever arise in the product. Whether justifiable or not, a microbiologist using nonstandard methods could be instantly discovered as the scapegoat by state prosecutors, by his own company, and even by many of his professional colleagues. The motivation to conform is strong.

He may even voice the fear of standing alone. It may figure prominently in his feasibility report; however, the emphasis on vulnerability will be shifted to the company. In reality, the relative merits of the proposed methods could be lengthily disputed — a sure sign of marginal superiority by either. Since the consequences are barely quantifiable, the degree of hazard assigned to them by any participant will depend very much on his motivation.

A research group has developed a rapid method of microbiological analysis yielding data correlating in 97.5 percent of the instances with the enumerative method for a variety of meats. Data have been published in scientific journals, presented at international conferences, and even a scientific instrument company has taken enough interest to manufacture apparatus for use in laboratories or factories. The Mk I instrument is a simple handheld probe, analogue readout, suitable for rapid checks around the factory. Mk II, intended to be the backbone of the line, is a more complex but very precise bench model with reagent pumps, TV monitor, and digital display. Mk III is an even more impressive full console model, interfaced to a time-sharing computer, fully programmable for automatic operation, or able to run in an interactive mode. Sales leaflets begin to arrive at hundreds of laboratories where the information in them becomes available to microbiologists. The same information is perceived by many eyes and apperceived in many different ways. Consider some of the possible reactions:

"I guess it does look like a scientific instrument. A bit awkward, perhaps, but then no doubt it was designed on a budget, for people on a budget. (Did the designer have to make it look more like a shoestring? Face it, we *are* the poor relations. . . .) Nice black knobs on a mass-produced bent steel case, neat and totally mundane metallic blue and beige. The usual glass/ceramic pumps clamped on here and there. . . . Lord, they look cheap and common! Can't someone produce something especially for microbiology so that it doesn't have to look like an afterthought? Even a little TV monitor with (believe it!) *three* different lighting positions and infinitely variable threshold (they mean I just have to hunt around the combinations until I get a result that looks about like what I already expected). And just look at some of these slogans:

ALLOWS THE TECHNICIAN TO RETAIN FULL TECHNICAL CONTROL

"They mean it'll pee all over the bench if I don't keep my eye on it, and the results won't be worth a dime to anyone unless I throw in my interpretation.

SAVES PRECIOUS BENCH SPACE

"Told you we were the poor relations.

OUTPUT EQUIVALENT TO THREE TECHNICIANS

"There are only three of us here anyway, and I'd go nuts here on my own.

ALL PREPARATORY AND WASHUP WORK ELIMINATED

"But Alice gets paid through corporate budget, not mine. . . .

97.5 PERCENT AGREEMENT WITH CONVENTIONAL METHODS OF ANALYSIS (Doe and Nobody, 1979)

"Did they invent it — or who was paying them? We'll have to work our asses off for weeks to get enough data to see whether we could trust it."

We could safely conclude that this scientist is unlikely to be motivated to go after the necessary dollars. He does not so much mistrust the apparatus as approach it with a mental set coloring his attitude to the entire concept. We can be fairly certain that if pushed to try the instrument in his laboratory, minor operational faults or discrepancies in its output would be seized upon to figure prominently in any report. In fact, it might be difficult to create a favorable image for the instrument. There are several points he could have picked up and amplified before his superiors into a strong argument for purchase. Could it be that he has a negative bias because past unfortunate experiences with microbiological instruments have dimmed his youthful optimism?

The impact on another scientist might be very different:

"Look at that design! No frills and money-wasting nonsense.

"Good old syringes! A bit sticky sometimes but pretty dependable, and the price is right if we break any.

"Enough variability in lighting and threshold to give me all the precision I need, and I won't have to relinquish my responsibility and training to a machine.

"We could get another water bath on the bench space we'd save. . . .

"We surely won't consider firing Bill and Greta, so we wouldn't save any money there, but maybe I could justify it on the extra testing and control we could do around the plant.

"Chemical QC Lab has been trying to get more of Alice's time anyway, so it would work out fine. . . ."

We could safely conclude that this scientist will try to buy an instrument. In reporting laboratory trials, the weights placed on operational problems and the significance of benefits or risks in using it would be quite different from those of the previous scientist. The price, appearance, convenience, efficiency, quoted data correlations, benefit/cost projections, etc., did not alter between the two outcomes. Only individual apperceptions of the information, the images generated in separate subconsciouses differed.

Had aspects of the instrument's scientific value, convenience, etc., been sufficiently strong, even the first scientist might have been swayed by the force of conscious logic. But how often does this occur in the field of microbiological analysis? Why was the second one biased so positively towards the instrument? Did something about its appearance or description excite the scientist subconsciously, evoking images of power, weaponry, masculinity, beauty, sexuality, intelligence, warmth, comfort, security, etc., if he/she possessed such an instrument?

a. An image of himself swaggering through the plant, probe swinging ominously from hip. "Hold 'er right there mister! I gotta probe here says this batch's contaminated 'n I ain't lettin' 'er through nohow! Hiya Burt! Let's take a look at thuh pork thet was worryin' yuh. Nope! Reckon she's OK! Let 'er through into Line B!"

b. An image of herself tensing behind the huge console, hair tossed back over her lab coat shoulder, beautiful face strained as with long manicured fingers she punches instructions into the great machine. A little circle of workers waits with bated breath, marvelling at the deftness of this delicate creature, her understanding and control. Lights flash. The printer is spewing paper. The young controller beams; a sigh of relief spreads through the crowd and across the plant. The result is good! The wieners can go out. . . . They are in business for yet another day. An old-timer touches her shoulder . . . "Bless you lass. . . ."

c. An image of a quiet but potent hum filling the room. The chairman's feet shift uneasily on the rich laboratory carpet.

"So you see, John, if you and your people could get us an answer on that pork shipment in the next twenty minutes, we could double this year's company profits . . ."

"Give us just fifteen minutes, sir, and I think we can come up with the information you need. Greta! Get this sample straight into Port B, hook in

Channels One, Four and Thirteen, and clear all lines so we can run straight through! And now, sir, you'd better leave. We're going to be pretty busy for a while. . . ."

Walter Mitty would have been proud to work with them. . . .

Fanciful? Certainly. Trite? Perhaps, for human motivations usually stem from much more complex sources than these simple daydreams. Insignificant in science? The conclusion lies with the reader. But let us go one step further.

The potential value of alternative microbiological techniques may be overlooked by the majority of scientists because of the scarcity of published corroborative data and favorable opinion. Since the number of publications is always less than the number of scientists acquainted with a new development, many scientists must try it out without being motivated to publish their interpretations. If it is accepted that orientation affects a scientist's apperception of the facility, advantages, aggravations, discrepancies, and other aspects of a scientific procedure, it requires only a small extrapolation to consider motivation to produce and publish convincing corroborative data as being similarly swayed. The generation of an unfavorable image will not motivate scientists to overlook early difficulties and obtain or communicate enthusiastic reports; in fact, it may even prompt the communication of destructive impressions. On the other hand, generation of a favorable image will motivate them to accept that early difficulties will be resolved and to produce data and arguments for its adoption.

In the production of supportive data or opinions, the true scientific merit of a development is of relatively minor importance (if, indeed, it can ever be known). Always, data and arguments for and against its acceptance can be found. If the development generates a sufficiently favorable image, those data indicating its advantages will be apperceived as being the most valuable, and there will be a motivation to communicate with this bias. Few want to be known against failing developments, but everyone would like their name associated with what may appear to be a universal technique of the future. Such a process, once started, can become self-catalysing, the image becoming increasingly favorable as approving publications come to the attention of still more scientists. Obviously, the scientific aspects must be "right." But this is often relatively easy to achieve; the problem is that they must be apperceived as right. In this, alternative microbiological methods have generally failed miserably.

If the idea of the associational content of data, methods and apparatus, and of conditioned responses being important to scientific judgement is objectionable, the message of this chapter will seem cavalier or irrelevant. If, however, it is agreed that scientific viewpoints can be colored by fears, desires, and experiences, the following conclusions will be seen to be important to the future of food microbiology:

1. While an enumerative basis for standards and methods may not be the most efficient way to control the wholesomeness of food, its survival has so far been ensured by psychological attributes not possessed by alternative approaches.
2. The enumerative approach to food microbiology will become increasingly untenable in the future, as its inhibitory effect on instrumentation and automation becomes more and more obvious.
3. The parameter of unwholesomeness approach may or may not provide the best framework on which a future food microbiology could be based. However, whatever the future basis may be, the efficiency of food microbiology in the future is certainly linked to the acceptance of nonenumerative methods. Its efficiency will depend, therefore, not only on the ability of instrument manufacturers to provide scientifically superior methods but also on their ability to exploit the same (or other) psychological factors.
4. Instruments or techniques satisfying these sufficiently will probably have little difficulty in catalyzing production of the scientific data necessary for their validation and general acceptance.

IMAGES OF BIOSCREEN

Whether or not the parameter of unwholesomeness approach personalized in BIOSCREEN (Chapter 5) is the proper replacement for enumerative microbiology, let us just examine it in the light of these conclusions. A determined effort has been made to remove the image of conventional scientific data in the initial information it conveys. This, of course, may be unpalatable to a few scientists. On the other hand, substitution of more generally emotive terms such as ALERT and OK (or CLEAR, ALL CLEAR, etc.) lends the results an immediate significance that is quite unusual to nonenumerative methods. In fact, the images projected by and the emotive impact of these terms may compete favorably with those generated by "numbers of germs."

Scientifically, the terms are quite precise, for they describe definite situations. They are instantly interpretable by scientists and nonscientists alike as infringement of or conformance to an agreed quality standard. To the nonscientist and for ACTION stimuli, these results are all that are important. The scientist, who will proceed to the second stage of perception — conscious interpretation and manipulation of the data — will find these primary data terms supported by BIOSPECS or BANESPECS containing the further information he/she requires. Scientifically also, of course, the data are unequivocal and much more defensible than numerical data, for they relate to a background of absolute data from which the standard was drawn.

From the regulatory point of view, the BIOSCREEN form of data may contribute positively to enforcement and encourage compliance. Its apperception by nonmicrobiological personnel in, say, factories as *light-flashing, tire-screeching emergencies* resulting from carelessness, or the *acknowledged safety and security* resulting from diligence would make it, I believe, a much more effective means of communication or feedback than even microbial numbers. The quality control microbiologist, for the same reasons, may find this data form contributing to an acceptance of sanitary principles around the plant.

The BIOSCREEN image should also be considered in the light of the instrumentation forms on which it may be obtained. Elimination of messy conventional operations such as blending, pipetting, and agar pouring, in favor of quiet, integrated incubator/detectors under minicomputer control, constant updates on the progress of analyses, and an interactive or conversational facility between operator and instrument may add an aura of scientific advancement and authenticity to the overall image that is barely possible with the enumerative approach. The instruments will be impressive and desirable. They will tend to reinforce the credibility of the microbiological data, thereby

Food Microbiology

motivating production of a rising spiral of supportive publications.

REFERENCES

1. Lowis, M. J.: The role of extract release volume in a rapid method for assessing the microbiological quality of pork and beef. *J Food Technol, 6:*415, 1971.
2. Shelef, L. A. and Jay, J. M.: Relationship between meat swelling, viscosity, extract release volume and water holding capacity in evaluating beef microbiological quality. *J Food Sci, 34:*532, 1969.
3. Kontou, K. S., Huyck, M. C. and Jay, J. M.: Relationship between sensory test scores, bacterial numbers and ERV on paired-raw and cooked ground beef from freshness to spoilage. *Food Technol, 20:*128, 1966.
4. Jay, J. M. and Kontou, K. S.: Evaluation of the extract release volume phenomenon as a rapid test for detecting spoilage in beef. *Appl Microbiol 12:*378, 1964.
5. Milne, R.: Floreat contagium vivum fluidum. *New Scientist, 77:*(1092) 567, 1978.
6. Dichter, E.: *Handbook of Consumer Motivations: the Psychology of the World of Objects.* New York, McGraw, 1964.

Chapter 8

A ROSE BY ANY OTHER NAME . . .

O UR PERCEPTION of microbiology today owes so much to the development of optical microscopes and nutritive agar jellies. An ability to see the tiny creatures involved in spoilage and disease was almost certainly *the* single facility leading to the rise of microbiology as a distinct branch of science. It seems to me also that once microorganisms had been identified as the causal factors, their apparent tangibility — the ease with which they could be handled, inspected, differentiated, described and named — inevitably drew them to the center of subsequent attention.

There is no doubt that in the early days of science, detection or quantitation of microbial presence was much easier using the fledgling counting methods than trying to determine any of their other manifestations. Biochemistry and immunology were as yet unnamed. Methods of physical or chemical analysis were too laborious, limits of detection too poor, and the number of recognizable parameters in any case so small that following the microbes themselves through their patterns of contamination, infection, and growth was the only sensible thing to do.

Since my aim throughout this book has been to suggest that maybe these days we focus more attention on the microorganisms than may be good for the science, it seemed a good idea to close by asking what might have happened had the optical microscope never been developed. Could the science of microbiology — or, at least, a similar science under another name — have grown up if we had never seen the central characters? Do we even require a *concept* of microorganisms in order to develop a science that is functionally very similar to the one we know? If that were so, how justifiable becomes the importance we observe for their role?

To answer this question, we need only consider some of the most obvious behaviors of microbial contamination, as we commonly describe them in the language of microbial numbers, and compare them with what we might have observed if we had been forced to rely on other scientific phenomena for our observations. It should be noted in passing that virology progressed for many years, developing analytical methods and valuable solutions to virological problems, before its fundamental particles could be seen. Today, the immunological and other manifestations of viruses remain much more important to their characterization and application than descriptions of the particles themselves. Consider the following summary of important microbiological features:

In some respects, microorganisms are quantized. The cell itself, for example, cannot operate without a minimum complement of functional structures, and colony-forming units behave as discrete particles. In other respects they are continuous. Glucose utilization, for example, is not quantized. Even the process of division is not a sharply defined jump from one to two cells, for the exact point at which it may be considered complete depends on the criteria by which one recognizes independent existences. Under suitable conditions, one unit may reproduce in a pattern of growth commonly described by three or four phases (Fig. 8-1A) in which their number per unit weight or volume of substrate may change several billionfold. Microbial cultures may be diluted and reinoculated to grow again. However, if they are diluted too much, regrowth becomes a matter of chance, depending on the probability of the inoculum volume containing at least one unit.

Rates of numerical increase depend on temperature, acidity, concentrations of oxygen and other nutrients, and on competition from other organisms. At very low temperatures, a rate of increase may become undetectably small, but the ability of the culture to reinoculate fresh substrate and multiply if conditions become favorable may be little diminished. Exposure to high temperatures may kill cells; their bodies may still be observable under the microscope, but they no longer possess the ability to multiply again, even when conditions should be very favorable. The rate of loss of virility frequently has a rather characteristic relation to the time/temperature exposure.

Figure 8-1. A. A growth curve — the commonly accepted representation of numbers of colony-forming units after microbial cells are inoculated into fresh culture medium. Following a readjustment period (Lag Phase), the count may increase exponentially until as much of the medium as possible has been converted to microbial cells (Maximal Population). An apparent loss of viability may then follow (Death Phase). B. Many other parameters related to microbial proliferation — particularly concentrations of structural molecules of cells, or metabolic products — exhibit similar curves. In the absence of a concept of microorganisms, growth curves similar to Figure 8-1A could still have been produced, but the nomenclature may have differed.

Cells suspended in diluent may be concentrated by centrifugation or filtration, subjected to chromatography, or precipitated with antibodies. Their ability to function may be inhibited or destroyed by extracts from other microorganisms and various chemicals. They may be transferred from hand to hand or in dust and many other vehicles. They may be identified coarsely by their characteristic shapes, mobility, patterns of gregariousness, and color, but more precisely by their infectivity, patterns of fermentation, chemical and physical by-products, and other biochemical or serological consequences. These characters are not immutable but may change with the nature of the environment and even from culture to subculture.

The significance of cells in spoilage or disease results from their collective and cumulative perturbation of the normal properties of the food or the host they infect, and these, in turn, depend on the characters just mentioned. Cultures and contaminants are quantified on the basis of cellular numbers or colony-forming units. However, cells possess their properties to very different extents and may also interact synergistically or antagonistically. Since many thousands of species, strains, and variants have been identified, describing probable hazards or the course of spoilage using only their numbers leaves us with many unknowns.

Now suppose that in some parallel world, development of analytical methods for detecting very small levels of many physical, chemical, and other properties had preceded development of the optical microscope and agar jellies. What might have been discovered?

Many parameters — physical phenomena, chemical substances and reactions — would have been discovered, each of which possessed the apparent property of self-amplification. For some chemicals, such as lipopolysaccharides, DNA, and ATP, apparent quantization would be observed, in the sense that certain minimum quantities or concentrations would be necessary before the property of self-amplification could be observed. Many other phenomena, such as glucose disappearance, proteolysis, and CO_2 evolution, would appear continuous. Under suitable conditions, these parameters would be found to change in a pattern describable by three or four phases (Fig. 6-1B) during which their amplitudes, e.g. concentrations, might change several billionfold. Parameters would apparently be capable of being diluted and reinoculated to grow again. However, if they were diluted too much, regrowth would be observed to become a matter of chance.

Rates of change of the parameters would be found to depend on temperature, acidity, concentrations of oxygen, and other relevant (passive) parameters. At very low temperatures, rates of change of amplitude might become undetectably small, but the ability of the culture to reinoculate fresh substrate and change again if conditions became more favorable might be little diminished. Exposure to high temperatures would appear to deactivate parameters. Sometimes the material, e.g. lipopolysaccharides or DNA, might still be detectable, but it would no longer possess the ability to self-amplify, even when conditions should have been very favorable. The proportion of a parameter losing its apparent propensity to self-amplification would have a rather characteristic relation to its time/temperature exposure.

These parameters could be concentrated by centrifugation or filtration; they could be separated by chromatography and precipitated by "anti-parameters" prepared from inoculated animals. Their ability to self-amplify would be found inhibitable by various chemicals or by extracts from the growth of other parameters. They would be transferable from hand to hand or in dust and many other vehicles.

All these aspects, along with many others that could be mentioned in the evolution of this science, sound rather familiar. But there would be differences:

Scientists in this other world would have no information about the shapes, mobility, and gregariousness of the cells giving rise to these parameters. Note, however, that the parameters by which they would grasp their science *are* exactly the quantified expressions of infectivity, contributions to patterns of fermentation, chemical and physical by-products, and other serological or biochemical consequences with which we in this world currently — but rather loosely — differentiate microorganisms. We are thereby forced to say that cells from different lines or in different environments have different activities. But variations in apparent "virility" of parameters with environment would have been a fact of life for this other science from its beginning. In it, however, the parameters themselves would be immutable. Many parameters *would be,* by definition, collective and cumulative perturbations of the normal properties of a food or infected host. A scientist measuring the proliferation of a parameter in spoilage or disease would automatically also measure synergism or antagonism. His data would, therefore, be of immediate and direct significance. A scientist studying the probable hazard from, say, a particular toxin within a food, might be blind to the scores or hundreds of species and strains in their various cell numbers, all interacting to control its generation. His analytical answer would simply tell the ability of that food to generate the toxin. What better answer could he obtain?

He might be surprised at the difficulties experienced by his counterpart in this world who, being unfortunate enough to have been raised with microscopes and agar, valiantly tries to guess the probable outcome of *so-many* cells of this species, interacting with *so-many* cells of that species, at *such-and-such* pH and at *whatever* water activity. He might wonder at the value of a science geared to this property called "cell numbers," which would seem to him to be so difficultly interpretable in terms of real live action.

It seems to me then, that many very similar phenomena and applications could have been observed had this science developed without the benefit of microscopes and agar. This alone seems to detract from the value of concerning ourselves too strongly with the properties of whole microbial cells. Moreover, had we never known about cells themselves, some of the most interesting manifestations of microorganisms — quantified in their ability to cause spoilage or infection — would inevitably have been based on parameters that were much more closely related to the point of interest than cell numbers.

This being the case, I would make the plea that while the century-old idea of the involvement of microbial cells has undoubtedly been a valuable stimulus to research, we might now look to a future where food microbiology is based on parameters more immediately relevant to the sensibilities of the consumer. The picture has changed over the years. At one time, only the classical analytical procedures of microbiology were capable of determining microbial presence; now, many other types of analytical procedures may have reached adequate levels of performance.

Almost certainly in the foreseeable progress of science, satisfactory methods will become available for determining most of the physiologically active parameters, provided we are satisfied to look for them at levels pertinent to human responses. The benefits from reorienting food microbiology towards the ability of foods to generate effective levels of these parameters would be in the formulation of standards defining specific hazards, facilitation of instrumentation, directness and unequivocality of data, improved enforcement, and speedier analyses, at least in urgent situations.

The effort required to accomplish this would be immense. Reorientations have occurred in other sciences, sometimes more than once. But food microbiology has enormously complex interests and pervasiveness across society; through its medical associations it would have the added burden of extreme caution.

At the least, however, it will do no harm to argue the idea from time to time. Few things but good come out of scientific controversies.

INDEX